SpringerBriefs in Statistics

More information about this series at http://www.springer.com/series/8921

Giuseppe Arbia

Statistics, New Empiricism and Society in the Era of Big Data

Springer

Giuseppe Arbia
Catholic University of the Sacred Heart
Rome, Italy

ISSN 2191-544X ISSN 2191-5458 (electronic)
SpringerBriefs in Statistics
ISBN 978-3-030-73029-1 ISBN 978-3-030-73030-7 (eBook)
https://doi.org/10.1007/978-3-030-73030-7

This Springer imprint is published by the registered company Springer Nature Switzerland AG
The registered company address is: Gewerbestrasse 11, 6330 Cham, Switzerland

Every century, a new technology - steam engine, electricity,

atomic energy or microprocessors –

has swept away the old world

with the vision of a new one.

Today it seems we are entering the era of Big Data"

Michael Coren

To my father, Francesco Arbia,

philosopher

Preface to the English Edition

This book largely represents the translation of an essay of mine published in Italian in 2018[1] with a couple of important differences. The first concerns the additional space which I devoted in this edition to the definition of a statistical sample design. During the Sars-Cov-2 pandemics, I was invited several times to national TV shows to help the public opinion interpreting the official data on the virus diffusion in Italy and to present my forecasts on its evolution. In these shows, most of the time my main job was dedicated to warning the audience towards incorrect interpretations of the data related to the number of infected people. In fact, those data were based on medical swabs that were not collected for statistical purposes, but only following criteria dictated by emergency and directed mainly to test the presence of the virus in people displaying severe symptoms. In many of these situations, I had to explain that, for this reason, the available data could not be used to estimate important pandemic parameters (such as the lethality rate and the fatality rate) at the level of the whole population. On that occasion, I realized that this aspect was not treated sufficiently in depth in the Italian version of my book and this suggested me to add in the present edition the entirely new additional Chapter 3.3 where I try to better clarify this aspect in an easy way accessible to a non-specialist reader. The second addition concerns the ethical issues which I discuss in Chapter 4. In the 2018 Italian version, I coined the term *Sixth Power* referring to the power of the Big Data within the context of the fundamental principle of the separation of powers in the theory elaborated by Montesquieu. In particular, in that chapter, I was pointing out the dangers connected with the power of handling large sets of personal information for commercial and political purpose and with the potential risks of interference with the other democratic powers. At the time, I was referring mainly to the famous scandal involving Cambridge Analytica and Facebook and of their supposed interference with the Brexit referendum and with the US 2016 presidential elections. However, in the two years which followed the first edition, new elements were added to the Cambridge Analytica-Facebook controversy and, just immediately before the book was submitted to the Publisher, the public opinion was again involved in lively discussions concerning big ethical

[1] See Arbia (2018).

questions related to the appropriateness of the social networks' managers intervention aiming at bunning single individuals from their platforms at their total discretion. Chapter 5 now contains a more updated account of the 2018 controversy and some considerations on the risk of non-independency between powers which may violate the principle of separation established by Montesquieu.

Like the first Italian edition, also this English edition is devoted to the memory of my dear father Francesco, who was a philosopher and filled my room with his philosophical books. Through those books, but so much more with his person and his passionate example of scholar, he passed on to me his *love for wisdom* (which is, indeed, the etymology of the word philosophy).

Rome, Italy
Epiphany 2021

Giuseppe Arbia

Contents

1 Introduction 1

2 What Do We Mean by the Term "Big Data"? 3
- 2.1 "Big" in the Sense of Large: The *Volume* 3
- 2.2 An Unceasing Flow of Data: Velocity 8
- 2.3 The World is Beautiful Because It Varies: The *Variety* 11
- 2.4 Problems Posed by Big Data: The Muscular Solution and the Cerebral Solution 12
- 2.5 A Fourth (and Last) "V": Veridicity 14
- 2.6 The New Oil: Big Data as Value Generators 15
- 2.7 Some Definitions of Big Data 18

3 Statistics and Empirical Knowledge 21
- 3.1 The Statistical-Quantitative Method: Measuring to Know 21
- 3.2 The Two Approaches to Reach Knowledge: Induction and Deduction 23
- 3.3 Obtaining Knowledge from a Partial Observation: *Good* Samples vs. *Bad* Samples 27
- 3.4 Some Epistemological Ideas: Inductivism, Falsificationism and Post-Positivism 36
- 3.5 The Lost Art of Simplicity: Ockham's Razor and the Role of Statistics 42

4 Big Data: The *Sixth Power* 45

5 Conclusions: Towards a New Empiricism? 51

Correction to: Statistics and Empirical Knowledge C1

References 53

Chapter 1
Introduction

Every century, a new technology - steam engine, electricity,
atomic energy or microprocessors –
has swept away the old world
with the vision of a new one
Today it seems we are entering the era of Big Data"
Michael Coren

Every day, in our daily actions, often unconsciously, we produce an immense amount of data. When we wake up, our smartwatch records our biological data, when going to work on the subway the obliterators trace our passage, while surfing the web we leave traces of our preferences and our IP address,[1] we withdraw the cash at the ATM that tracks the operation, if we have a sports activity an *app* records our performance, while shopping we use a loyalty card which records our preferences. Many other operations leave daily an indelible trace of our activities. Every day, moreover, in order to take our decisions (small or big), we make continuous use of a set of data recorded by others. In choosing a trip we compare the prices of different tour operators, before making a financial investment we read the stock market trends of various financial products in recent years, in order to decide where to spend an evening at dinner we read on our smartphone the scores of various restaurants trying to choose the one that suits most our needs in terms of the quality, location or price viewpoint.

Nothing radically new! For centuries, mankind has tried to take its own decisions based on the information available. However, never before have information come to us with such rhythm, variety and volume as they do nowadays. This huge flow of data that we produce and to which we access daily (and whose volume grows at an accelerated speed over the years) substantiates into a series of numbers, texts,

[1] The IP address (Internet Protocol address) is a numerical label that uniquely identifies an electronic device connected to a computer network via the Internet.

G. Arbia, *Statistics, New Empiricism and Society in the Era of Big Data*,
SpringerBriefs in Statistics,
https://doi.org/10.1007/978-3-030-73030-7_1

images, sounds, videos and more, which has the potential to change radically the way in which we take our decisions as individuals, enterprises, public administrations and as society in all fields of human action.

We refer to this phenomenon with a term which has forcefully entered into our common lexicon: *Big Data*!

The purpose of the present essay is manifold. First of all, given the confusion and ambiguity that is still around this term, we aim at defining in a precise (although absolutely non-technical) way, what we mean with Big Data clarifying its peculiar characteristics and trying to make the topic as accessible as possible to the non-specialists. Secondly, we aim to discuss the critical issues and problems which are related to the phenomenon of Big Data and their possible consequences in everyday life.

Finally, we aim to discuss the consequences of the Big Data revolution on Statistics: the art of knowing reality and taking decisions based on empirically observed data.

Chapter 2
What Do We Mean by the Term "Big Data"?

2.1 "Big" in the Sense of Large: The *Volume*

In this essay, we will speak about data and about their enormous quantity and variety that is overwhelming us in these years.

Only until a few years ago, there were only a few experts (statisticians and information technology people above all) who had to deal with this topic, while today everybody continuously performs operations that generate data and, on the other hand, always takes decisions based on them. Since this phenomenon tends to grow over time at an accelerated rate, the subject is becoming increasingly common.

In dealing with data, it is useful to introduce some concepts related to their measurement in order to quantify the volume and to make the reader more knowledgeable about its recent growth and explosion.

For centuries, data and information that were not transmitted orally could be quantified in terms of the pages (collected in papyri at first and books later) and in terms of the libraries that contained large quantities of them.

The Library of Alexandria in Egypt, the largest and richest library in the ancient world founded around 300 BC, contained scrolls, codices and papyrus summarizing much of the knowledge available of the time. It is estimated that the scrolls were about 500,000, and all of them were (alas!) destroyed by the Romans first and then by the Arabs. After the invention of printing and the consequent increase in book production, the library becomes the symbol of the vastness of human knowledge. The United States Library of Congress in Washington, for example, currently represents one of the most complete collections of knowledge and contains, to date, about 158,000,000 volumes.

Things started to change dramatically at the beginning of the 1940s of the last century when, with the advent of information technologies, data were assigned a unit represented by the so-called bit (acronym of binary digit),[1] through which every

[1] A measure introduced by Claude Shannon (1948).

G. Arbia, *Statistics, New Empiricism and Society in the Era of Big Data*,
SpringerBriefs in Statistics,
https://doi.org/10.1007/978-3-030-73030-7_2

piece of information was reduced to a sequence of binary symbols: the digit 0 and the digit 1. In this way, two bits allow to represent four sequences (i.e. the sequences 00, 01, 10 and 11), three bits indicate 8 distinct sequences (000, 001, 010, 011, 100, 101, 110 and 111) and so on for increasing values of bits. In particular, eight bits allow representing 256 different sequences (a number given by 2 raised the power of 8) which indicate an additional unit, a multiple of the bit, which has become popular and has now entered the common language: the *byte*.

Each of us has familiarity with this unit in our ordinary life. The older among the readers certainly remember the first and rudimentary computers that entered our houses in the 1980s of the last century. These computers, at the beginning, had a mass memory with a capacity of 65,536 bytes (which we can indicate as 64 kilobytes, or 64 Kb, making use of another unit of measurement) as, for example, the celebrated Commodore 64. Towards the end of the decade, these computers could store up to 640 kilobytes. Moving on in time in this short history, the personal computers at the beginning of the 1990s have been equipped with a fixed memory which was of the order of 10 megabytes—the megabytes being a further unit of measurement that corresponds to 10 million bytes. Nowadays, we carry with us in our pockets smartphone devices which can store data (phone numbers, images, movies, music and many others) which occupy a storage space of the order of 256 gigabytes (or 256 billion bytes), while our laptops are endowed with a memory that often reaches one or more terabytes (i.e. a measurement unit that corresponds to one thousand gigabytes or, in other words, one trillion bytes!).

These figures, which now seem impressive, will make the reader of this essay smile only in about ten years or so assuming optimistically that there will be still someone interested in such an argument that will, most probably, appear absolutely obsolete.

To better understand the dramatic increase in the volume of the available data, in a work entitled "*How much information*", Hal Varian and Peter Lyman (two economists from Berkeley University) calculated that the total production of data at a world level in the year 2000 amounted to about 1.5 exabytes (or 1.5 billion gigabytes) about 37,000 times the aforementioned Library of Congress of the United States. Three years later, the same researchers repeated the same calculation, estimating that the volume had increased to 5 exabytes with a growth of 66% per year over the period considered!

These are mind-boggling figures!

In 2003 alone, therefore, 5 billion gigabytes (referring to the unit of measurement we are more familiar with) were generated!

In the last decades, the evolution of the volume of data has been exponential: in 1986, the data produced were 281 petabytes; in 1993, they increased to 471 petabytes; in 2000, they became 2.2 exabytes; in 2007, they were 65 exabytes; in 2012, they reached 650 exabytes.[2]

However, there is an even more striking quote that helps the reader to quantify the explosion in data production in the last decades. In 2010, Google's CEO, Eric

[2]See McAfee e Brynjolfsson (2012).

Schmidt, stated: "*There were 5 Exabytes of information created between the dawn of civilization through 2003, but that much information is now created every 2 days*" (Schmidt, 2010). Schmidt's forecast, indeed, proved to be an underestimation. In 2013, we had already produced 4,400 exabytes, while following the calculations of Schmidt we should have accumulated only 3,000. In 2021, we expect to produce 44,000 exabytes.

For such quantities, it is necessary to introduce a further unit: the *zettabyte* which corresponds to 1000 exabytes. The International Data Corporation (IDC)[3] predicts the world's data will grow to 175 zettabytes in 2025.[4] We cannot even imagine such quantities. Just to have an idea, if I have to store 175 zettabytes on DVDs, the stack of DVDs would cover the Earth's circle 222 times and it would take you 1.8 billion years to download them at the current Internet connection speed.

For these quantities, we will soon need another unit of measurement which has already have defined: the *Yottabyte* which corresponds to a thousand zettabytes.[5]

The creation of new terms for the measurement units proceeds as fast as the increase in the amount of data!

The value estimated by Schmidt and all the previous examples are much less impressive, however, if we consider the content of the data that are generated daily and we start making an important distinction between data and information. In fact, it should be kept in mind that most of these data is substantially irrelevant from the viewpoint of information and of knowledge building and transmission. In fact, in the 5 exabytes generated "*from the dawn of civilization until 2003*" calculated by Schmidt, there are books such as the Bible, the Koran, the Divine Comedy and Beethoven's Ninth Symphony. A document containing the Bible can be stored in a file of about 4.13 megabytes, about the same amount required by the picture I'm shooting with a medium-quality smartphone at the fish dish I'm eating in order to show it to my friends to make them gnawing from envy. When I then share the same photo with 20 chat friends, in a second I have produced the same volume of data that corresponds to 20 times the Bible! With an information content which I leave to the reader's judgement!

Here is the first (perhaps not so obvious) feature to keep in mind when dealing with Big Data:

Accumulating a lot of data does not necessarily increase the information available for the knowledge of the phenomena and to assist our choices in taking decisions.

There is indeed a substantial difference between "data" and "information".

Part of the ambiguity in the use of the two terms is certainly attributable to the use we currently make of the word "informatics". This term was introduced by Philippe Dreyfus in 1962, by the contraction of the French terms informat(ion) (automat)ique,

---

[3]https://www.idc.com/about.

[4]See https://www.seagate.com/gb/en/our-story/data-age-2025/.

[5]For clarity, we summarize here the various units of measurement we refer to in the text. Kilobytes (one thousand bytes), megabytes (one million bytes), gigabyte (one billion bytes), terabytes (one thousand gigabytes), petabyte (one million gigabytes), exabyte (one billion gigabytes), zettabyte (one thousand exabytes), yottabyte (one million exabytes).

and independently by Walter F. Bauer and associates who co-founded a software company called *Informatics Inc.* Indeed, the new discipline that was being born should have been more correctly Christianized with the term *Data-matics*, to make it clear that it refers to the data and not per se to the information contained therein.

The distinction between data and information is very well expressed by the popular science fiction author, Daniel Keys Moran, when he states that: "We can have data without information, but you cannot have information without data".

I cannot find, however, any better way to deepen this theme than the novel "*The Library of Babel*" written by Jorge Luis Borges.[6] In this novel (of which I will report an excerpt below), the Argentine writer starts from the following intuition. A book is made up of, say, 410 pages, each of which contains 40 lines where each line is a sequence of 40 characters. Therefore, we can think of a book as a collection of something less than 700,000 characters.[7] These characters can assume only a limited number of possible values: the letters of the alphabet, the space and some punctuation symbols (the dot, the comma, the colon, the semicolon, etc.). Let's say that these characters are globally 30. If we consider all the possible combinations of these 30 characters in the 410 pages of the book, we are considering all the books that can be conceived. The number of these books is very large, although not infinite.[8] Since the number of all possible books is finite, in his novel Borges imagines that there is a library (which he fantasizes to be architectonically structured into large adjacent hexagons) which can contain all of them. Here is what the author tells us, taking us by the hand in this suggestive place of fantasy.

> *This thinker observed that all the books, no matter how diverse they might be, are made up of the same elements: the space, the period, the comma, the twenty-two letters of the alphabet. He also alleged a fact which travellers have confirmed: In the vast Library there are no two identical books. From these two incontrovertible premises he deduced that the Library is total and that its shelves register all the possible combinations of the twenty-odd orthographical symbols (a number which, though extremely vast, is not infinite) that is, everything it is given to express: in all languages. Everything: the minutely detailed history of the future, the archangels' autobiographies, the faithful catalogues of the Library, thousands and thousands of false catalogues, the demonstration of the fallacy of those catalogues, the demonstration of the fallacy of the true catalogue, the Gnostic gospel of Basilides, the commentary on that gospel, the commentary on the commentary on that gospel, the true story of your death, the translation of every book in all languages, the interpolations of every book in all books, the treatise that Bede could have written (and did not) about the mythology of the Saxons, the lost works of Tacitus.*
>
> *When it was proclaimed that the Library contained all books, the first impression was one of extravagant happiness. All men felt themselves to be the masters of an intact and secret treasure. There was no personal or world problem whose eloquent solution did not exist in some hexagon. The universe was justified, the universe suddenly usurped the unlimited*

[6]It first appeared in 1941 in the collection of novels entitled *The garden of forking paths* and then in 1944 in the volume *Fictions*. See the translation in Borges (1999).

[7]A number derived from the product of 410 pages times 40 rows per page, times 04 characters per row which equals a total of 656,000 characters.

[8]For the lovers of combinatorial calculus, this number is given by 30 raised to the power of 656,000, but if we try to calculate the exact number, any computer software will not provide an answer and will only tell you: "infinite"!

dimensions of hope. At that time a great deal was said about the Vindications: books of apology and prophecy which vindicated for all time the acts of every man in the universe and retained prodigious arcana for his future. Thousands of the greedy abandoned their sweet native hexagons and rushed up the stairways, urged on by the vain intention of finding their Vindication. These pilgrims disputed in the narrow corridors, proffered dark curses, strangled each other on the divine stairways, flung the deceptive books into the air shafts, met their death cast down in a similar fashion by the inhabitants of remote regions. Others went mad ... The Vindications exist (I have seen two which refer to persons of the future, to persons who are perhaps not imaginary) but the searchers did not remember that the possibility of a man's finding his Vindication, or some treacherous variation thereof, can be computed as zero.

I think it was worthwhile to report this long quotation of the novel. First of all because it is fascinating in itself, but also because I wanted to draw reader's attention to the fact that many data do not imply necessarily more knowledge. On the contrary, having available all the possible sources of information (all that is given to know about past, present and future events), but without having a precise key for understanding (in the symbolism of the novel of Borges, this key would be a map of the library, a faithful catalogue), reduces the possibility of reaching true knowledge to zero. Let's think about it. Indeed, many volumes of the library are just constituted by blank pages with only some nonsensical characters scattered here and there in the empty pages. Furthermore, the overwhelming majority of these books is just a collection of nonsense sequence of characters and words. Among the few volumes that contain sequence of words with a sense, there is still a vast majority of books reporting misleading realities that do not correspond to the truth. Only an imperceptible minority of books are describing the true reality and are useful to increase our knowledge. (Hopefully this book you are now reading belongs to that small subset!).

This is exactly the situation in which we currently are when we observe the explosion of Big Data.

We have a huge amount of data, an incredibly large Babel library, tending to infinity, but many of these data (indeed the vast majority of them!) are deceptive, contain errors and incompleteness or are absolutely irrelevant to acquire knowledge about the phenomena we are interested in. Indeed, they could be trashed without any relevant loss from the point of view of building a greater knowledge that is useful to support our individual and collective choices.

To explore the Babel of big data without running the risk of getting lost, we need the non-deceptive catalogue, we need a methodological approach, and we need a "*method*".

We will go back to this topic in Chapter 3.

The increasing amount of data available requires always larger computer memory. Therefore, the continuous growth of the data at our disposal needs engineers and technicians striving to answer to the increased demand by introducing new equipment with an ever-increasing memory that is capable to store and process them.

Babel's library always needs new buildings and new hexagons to store the new volumes that we generate on its shelves.

2.2 An Unceasing Flow of Data: Velocity

The large volume of data currently available is certainly the most obvious feature of Big Data, but absolutely not the only one. In many applications, in fact, the speed of data generation is even more relevant than its volume and the ability to treat it is more challenging. Indeed, not only do data accumulate creating problems of storage, management and interpretation, but (like everything else in our contemporary society) they travel at a speed which itself poses previously unknown problems of collection, storage and analysis. All data that we unconsciously produce on a daily basis with our routine operations are increasingly transmitted substantially in real time. Consider, for instance, the data related to millions of bank customers who perform their operations online or at ATMs or by credit card or cashless payments from their smartphone. These data accumulate and reach the operators of the banking sector substantially in real time. In a similar way, the *likes* or *posts* or any comment reported on the social networks can be collected continuously over time as well as, e.g., the record of our sports performances, the environmental data collected by monitoring stations, the data related to urban traffic and many others. Moving to a different example, in the healthcare sector, data related to diagnostic tests and medical measurements (e.g. blood pressure, temperature, etc.) can now be collected in real time on each patient. In future, these pieces of information will be added with other data (such as treatment, guidelines, electronic medical records, doctor's notes, research materials, clinical studies, journal articles and personal information on the patient), constituting a real-time support system for clinical decisions that will work at the speed of light.

These are just a few examples. Many others can be found in every sphere of human activity (related to individuals, companies and public institutions) and will increasingly enable us to change substantially the process of decision-making treating streams of data at unimaginable speed.

Let's just think about what could be the incredible advantages that the ability to collect and process data at the speed of light can provide in many different situations. A dramatic, but very good example of this is the case of an epidemic surveillance, monitoring and control like the recent *SARS-CoV-2* pandemic of 2020. In this case, if news about newly infected people could be obtained in real time and monitored in a continuous space, the spatial diffusion of the virus could be monitored, so that the outbreaks could delimited and kept under control. A further example is constituted by computer-assisted surgery, where, due to our increasing ability to collect data through medical imagery and instruments, a surgeon can view dynamic 3D images of the affected part while recording the patient data in real time together with his/her reactions during the surgery. Other possible examples could include environmental protection (in the prediction and real-time control of natural disasters), crime prevention, meteorology and so many other fields that are impossible to enumerate exhaustively.

In economics, for instance, the possibility of collecting data in near real time gives the obvious competitive advantage of providing a company with the possibility of being more agile and resourceful than its competitors, of anticipating their policies thus gaining profit by taking strategic decisions.

If it is certainly true that the availability of data at the speed of light opens up new perspectives in very different fields, on the other hand, likewise the volume, the increasing speed of data collection also raises new computational problems which need adequate hardware structures and software codes to be solved.

In the same way, in which technological evolution has tried to cope with the increased need for data storage endowing computers with larger and larger memories, the growing dynamics with which data are generated also require an evolution in the speed with which collection, storage, management and calculation operations can be performed.

The calculation speed, like the volume of data, has its unit of measurement. Indeed, a computer processor is made of various logic circuits, which are responsible for carrying out various operations and which interact with each other to exchange information. The calculation speed refers to the number of elementary operations that circuits are able to execute within one second. This defines the unit of measurement called *Hertz* (indicated with symbol *Hz*) named after Heinrich Rudolf Hertz, the German physicist who discovered the existence of the electromagnetic waves.

The more Hertz a computer is equipped with, the greater is the number of operations that can be performed within the same time frame.

One of the first computers, which used electromechanical technology (the famous Z1 built by Konrad Zuse in 1938) had a speed that reached a maximum of 1 Hz. The speed of the first modern electronic microprocessor, the pioneering Intel 4004 designed by the Italian engineer Federico Faggin in 1971, was working at a speed of around 740 *Kilohertz* (740 thousand of Hertz) that is 740,000 operations per second. This processor has evolved over the years into increasingly faster machines (the models 4040, 8008, 8080 and 8085), progressively reaching in the mid-1970s a speed of the order of 2 Megahertz (or *MHz*, for short, corresponding to 2 million operations each second). In this short history, someone may remember the first home computer, which was equipped with processors (the famous 8086 to 80,486) and was working at a speed that could reach 50 Megahertz in the mid-1990s. In more recent years, we have become accustomed to a further unit of measurement: the *Gigahertz* (GHz) corresponding to the possibility of performing one billion logical operations in one second. The advanced Pentium III "Coppermine" processor was the first to overcome the threshold of 1 Gigahertz—something that was absolutely unconceivable only until a few years earlier.[9]

The processors installed on our current personal computers now can reach a speed of around 8 GHz. Figure 2.1 summarizes graphically the evolution of computer capability to deal with several operations in a unit of time.

Two problems emerge in this continuous rush towards faster processors. Firstly, the calculation speed is limited by the slowest logic circuit. It may therefore happen that some very complex calculations significantly reduce the overall performance of the whole system. Secondly, modern processors have now reached the maximum

[9] https://www.tomshardware.com/reviews/intel-cpu-history,1986-9.html

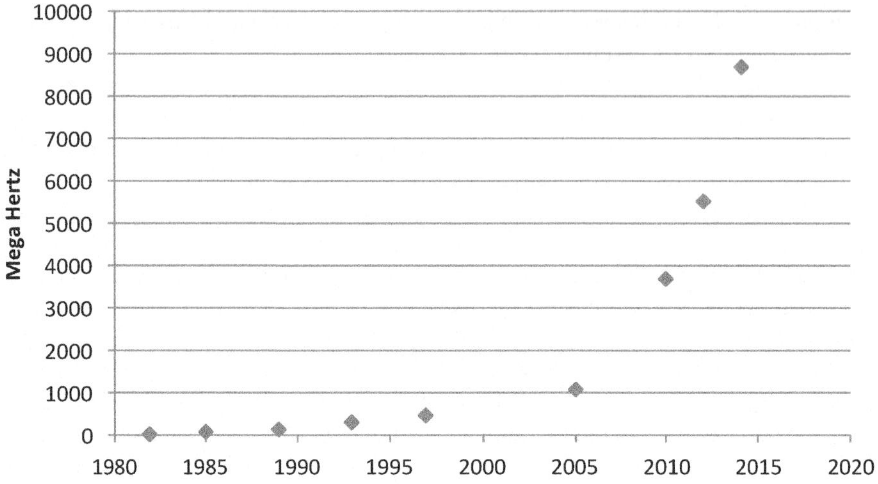

Fig. 2.1 Evolution of the speed of processing in the period 1980–2015

speed allowed by the current technologies. Above this speed, in fact, the heat produced during processing can no longer be disposed of and can cause damages to the processor. To overcome both problems, waiting for more satisfactory technical solutions, we try to reduce the maximum length of the logic circuits by distributing the most demanding calculations on different processing units. It is the idea of *parallel* (or distributed) *computing*. The technique called "parallelism" has long been used with large supercomputers, but it has received a wider interest only recently due to the physical constraints connected with heat generation so that, in recent years, it has become the dominant paradigm especially in the form of multi-core processors also for personal computers.[10]

In its essence, parallel computing refers to a typology of calculations where the many operations involved in their execution can be carried out simultaneously. Indeed, large problems can often be subdivided into smaller ones, and the calculation can be run at the same time. This enhances enormously computer performances when dealing with very large and quickly increasing datasets. Indeed, with Big Data of the order of zettabytes (billions of terabytes), the distributed computing power requires a massive effort with dedicated tools running on hundreds and sometimes thousands of computer servers working in parallel.

Once again in the presence of growing speed of data collection, we answer by producing new models of computers that are able to keep up with it. As a consequence, even in this case, like in the case of the volume, however, we cannot see the end of this race, at least in that brief horizon of time we can look at, nor can we predict accurately if we will reach a limit in the physical possibility of data collection and storage.

[10] A multi-core processor is a processor on a single integrated circuit which is endowed with two or more separate processing units.

2.3 The World is Beautiful Because It Varies: The *Variety*

In the previous chapters, we have identified two fundamental characteristics of Big Data: their volume and the velocity with which they are acquired. With a definition that has become famous and cited in every scientific forum, an analyst of the Gartner company, Doug Laney, in 2001 identified a third important feature of Big Data: the Variety. Being also characterized by the initial "V", he thus provides a description of the phenomenon of Big Data which became famous as *the 3V's definition.*[11]

Indeed, before the advent of Big Data, analysts have become used to collecting quantitative or qualitative data stored in tables and ordered into rows and columns using tools like, for instance, the *Excel* spreadsheet of the *Office* package that many of us know. Everyone involved in data analysis, or even just in their organization and management, has been familiar for years exclusively with this type of data. Data ordered in this kind of tables can be easily synthesized using simple descriptive procedures (such as the calculation of averages and percentages and the production of graphs and tables, just to mention those that constitute a common knowledge even among non-specialists statisticians) and then used for more ambitious purposes using sophisticated models that constitute the statistical toolkits available for forecasting and decision-making. Today, we refer to this type of data as "structured data".

In more recent years, the explosion of the Big Data phenomenon has increasingly brought to the attention of analysts new data coming from very heterogeneous and unconventional sources, which cannot be stored in the simple structured databases used until now. In fact, the sources of data are now no more limited to archives, administrative records, survey or panels as it was in the near past and they more and more include alternative sources such as satellite and aerial photographs, images, information obtained through drones, GPS data, crowdsourcing,[12] cell phones, web scraping, Internet of Things (IoT) and many others. In addition, we can also observe and collect a plethora of other non-traditional data like, e.g., emails, *WhatsApp* texts, tweets or other messages, video or audio clips, text data, information taken from social networks (posts, likes, comments, pokes), multimedia contents and many others. We refer to this wide class of new data as "unstructured data". The greater the complexity of data, the more difficult becomes the task of storing them, managing them into databases, linking them together and obtaining interesting information from them.

According to an IBM estimate appeared on the Big Data & Analytics Hub website,[13] unstructured data currently nowadays account for 80% of the data and only the remaining 20% is represented by the more traditional structured data.

The availability of unstructured data of very different varieties has constituted a discontinuity break with respect to the past even more than the volume and the speed with which they are collected. Indeed, the variety of the data, even more than their volume and their speed, requires a radical revision not only of the collection and

---

[11] See Laney (2001).

[12] A term derived from the merging of the words *crowd* and *sourcing*.

[13] https://www.ibmbigdatahub.com/infographic/big-data-healthcare-tapping-new-insight-save-lives.

storage procedures, but also of the methodologies of analysis. In particular, from the viewpoint of collection and management due to the variety of unstructured Big Data, traditional relational databases are totally inadequate and do not allow to archive and manage them at an adequate speed. This has led researchers to look for alternatives which make storage and subsequent operations more agile, by preferring those that are now referred to as NoSQL (Not Only Structured Query Language) systems. The elective tool among these unstructured systems is the software called "Hadoop", a program designed for the storage and processing of Big Data which can deal with distributed operations with a high level of data accessibility. Its creator, Doug Cutting of the Apache Software Foundation, in 2003 was working on a project aimed at building a search engine, when some unforeseen difficulties arose related to the vastness and variety of information that the system had to manage. In order to solve the problem, he created a completely new product to which he gave the name of his son's favourite toy: a yellow rag elephant named Hadoop.

Hadoop nowadays allows us to carry out operations that were intractable until a few years ago, by employing the structured databases we were used to. However, it should be noticed that, once again, the approach we use to deal with problems related to the new variety of information gathered through Big Data is to try and adopt the same approach employed up to now, to the changed needs.

We strongly believe, conversely, that we now need a discontinuity and a revolutionary step in the way we approach a situation which is totally new.

2.4 Problems Posed by Big Data: The Muscular Solution and the Cerebral Solution

We have seen in the previous sections that the spread of Big Data characterized by immense volumes, uncontainable speed and always new varieties, requires a continuous evolution of electronic instruments in terms of their memory, processing speed and access modes, with a spiral of which it is not possible to see an end.

If we think about it, the phase we are experiencing is absolutely not new in the history of humanity. Indeed, men have continuously faced epochal changes, facing problems that seemed insurmountable, which required a break from the past, a change of attitude and a search for brand new solutions.

In the Neolithic period, for example, men started devoting themselves to agriculture and pastoralism, thus passing from a nomadic to a more sedentary stadium. In this process, they needed to abandon the caves and huts that had offered them a good shelter for several millennia until then and began to build the first houses using large wooden beams, stones and the first bricks. In this new situation, the alternatives they had available were two. The first consisted of keeping on doing what they had done up to that moment by simply modifying the scale of their intervention. If until then a construction (such as a hut) required the effort of only one or few individuals, to build houses in stones and bricks it was necessary to multiply the efforts and therefore involved a larger number of people. The first solution therefore consisted only of multiplying the muscular strength employed. An alternative solution, much more promising, however, consisted in radically changing the approach to the problem thus introducing technical innovations (such as the slide, the wheel and subsequently winches and other construction tools) which made it possible to move weights heavier than those considered up to that moment and to overcome the new and increased difficulties. It should be noted that this second solution, while solving that specific practical problem, opened up new and absolutely unthinkable, possibilities. Surely our ancestors, the cave men who loaded a large trunk on a cart endowed with rudimentary wooden wheels, did not imagine that the same instrument could in the future allow people and goods to be moved at 300 and more kilometres per hour!

In the case of Big Data, the situation is exactly the same in that the solutions to the challenges posed by their uncontrolled spread may be of two kinds. The first consists in continuing to do what we have done up to now, considering it essentially a problem of adapting the information technologies and, therefore, concentrating our efforts in the design and implementation of increasingly powerful computing systems in terms of manageable memory, speed of calculation and versatility of access. This is the solution we are experiencing these days when we often perceive the Big Data problem only as something that has to be faced by computer engineers. The second consists in accepting the current challenge by radically changing our approach and, therefore, in imagining new methodologies which could allow us to get out of the spiral in which we are currently locked up and, while indicating a way out of the current problems, could suggest new (and unimaginable) possibilities of future applications.

The first approach could be referred to as *evolutionary* in contrast to the second that we could call *revolutionary*. However, we prefer (in homage to the example we have reported at the beginning of this section) to call the first solution a *muscular solution* and the second a *cerebral solution*. The former is easier and more convenient, but it is doomed to fail in the long run. The latter is more difficult because it requires an effort of imagination and abandoning established practices. I want to emphasize that we often face such choices even in solving our small everyday problems. In this respect, I remember an episode that occurred to me long ago. I was a guest of a friend who had a small swimming pool in his country house. When I visited him, he had the problem that the cover sheet of the pool was filled with water as a result of a violent storm and it was likely to tear off. Thus, upon my arrival, I found him in the process of trying to remove the water from the tarpaulin with a bucket. Having a personal distortion towards a quantitative approach to problems, with a quick calculation I reached the conclusion that it would have taken him about 12 h to empty the cover sheet completely from water. With an excuse, I went away and returned after an hour (finding him still intent in his work) with a cheap electric pump, with which I emptied the canvas in about half an hour. Months later, he told me that the cellar of his house had flooded and, having available the electric pump that I had bought, he could quickly solve the problem. The brain solution forces us to change our approach, and our comfortable habits, but it is always the solution to be preferred. And it could help solving not only the present challenges, but also future unforeseeable problems.

The way of dealing with the problems posed by the management and analysis of Big Data is not different. The solution cannot only be computer-based: it must necessarily be grounded on the joint use of computer science and the statistical method. Indeed, statistics represents the method to approach reality, based on empirical data, which tends to quantifying and simplifying it in order to produce empirically founded decisions: a method that for centuries has informed human knowledge in all fields.

We will devote our attention to this important aspect in Chapter 3.

2.5 A Fourth (and Last) "V": Veridicity

The characterization of Big Data based on the 3 "V's" introduced by Doug Laney dominated all the textbooks and articles devoted to the topic since the beginning of the millennium. In more recent years, however, the discussion on the subject has led to identifying a fourth element which is typical of the phenomenon and it is also characterized by the initial "V": the one related to veridicity.[14]

Indeed, considering the great variety of data sources (structured and unstructured) that are continuously collected, very often it is not possible to keep the quality under control and they can be contaminated by errors and inaccuracies of various kinds that are of a higher complexity than those usually encountered in the traditional processes of data collection. Obviously, if the basic empirical data are not sufficiently accurate, the results of the analyses will not be reliable. Statisticians describe this aspect with the motto: "garbage in, garbage out". Since important decisions are usually based on these analyses, it becomes of paramount importance to be able to measure data veridicity and reliability and try to clean the data from their imperfections prior to any form of analysis in order not to run in gross errors and dramatic consequences in our decisions.

There are several inaccuracies that emerge in the collection of Big Data and this is certainly not the place where the topic can be treated with thoroughness and rigour. However, looking at the most obvious aspects, we can observe, for instance, that in gathering data through social networks, there are very well-grounded doubts about their truthfulness, since often those who provide information on their own profile have an interest in supplying them incorrectly. Moreover, often the identity of the user can be wrong or even refer to non-existent people and often individuals can have duplications or even multiplications of their personal profile. Furthermore, the judgements, facts and comments reported on social media can be intentionally falsified in order to provide a biased image so as to condition decisions and choices of other users. This is the case of the so-called *fake news* which, due to the mere fact of appearing on a social network, acquire some resemblance of officiality and truthfulness despite their validity. The use made of this type of news in electoral events or in commercial promotion activities is well known and will be treated to a further extent in Chapter 4.

---

[14]https://www.ibmbigdatahub.com/infographic/four-vs-big-data.

It is not only in the social media that we can find inaccuracies of this nature. For instance, important sources of Big Data are the already mentioned crowdsourced data, that is, data collected by a large public in certain contexts on a voluntary basis. This typology includes, for instance, data related to health symptoms collected through cell phones and employed in a system of epidemic surveillance. A similar example relates to car traffic data transmitted from a device to a central collection point in order to report road conditions, accidents, traffic jams, work in progress and other related information. But we can have many more examples relating to seismic events, natural disasters, food prices in developing areas and other situations in which traditional data harvesting criteria are more difficult, slower or even impossible. Although these sources of data are potentially very valuable, their veridicity cannot be assumed uncritically since the observed phenomena can be over- or under-estimated due to errors made (intentionally or not) by those who provide them on a voluntary basis.

On more technical, and perhaps less intuitive level, errors that reduce (and in many cases nullify), the usability of Big Data can arise, for example, from the inaccuracy of measuring instruments (e.g. in radiological images, from a microscope or from satellite), from the incompleteness deriving from cancellations or missing data on administrative registers, and from data which are collected without following a precise statistical sampling criterion. This is an aspect that deserves a more thorough discussion which we will present in the following Sect. 3.2.

Furthermore, in some cases, the data available are intentionally modified in order to protect the individuals' privacy. This is the case, for instance, of many health information (such as the presence of a certain pathology in a patient), which can be collected with a high precision recording the precise geographical coordinates of the patient, but is then modified in order to prevent the individuals to be traced with certainty, thus violating their privacy.[15] Such interventions, although appropriate for the reasons of confidentiality mentioned above, can in some cases radically modify the reliability of the set of data analysed and dramatically distort the information up to a point which can lead to incorrect decisions.

These are all aspects of data veridicity on which Statistics has been questioned in recent years, trying both to highlight the reliability of a datum through appropriate indices and to develop new techniques for cleaning and correcting the data in order to reduce the distorting effects on any subsequent analysis.

2.6 The New Oil: Big Data as Value Generators

Having discussed at length the main characteristics of the Big Data, it should now be clear why many see in them the new oil.[16]

---

[15] An operation called *geo-masking*.

[16] "The data is the new oil" says Ann. L. Winblad, of Hummer Winblad Venture Partners. However, someone says: "Data is not new oil" (Jerp Thort, Data artist and columnist of the New York Times).

Only until a few years ago, companies stored daily data in their traditional databases through which they gathered information about their operations with no ambition to use them for different scopes. Today, it would be impossible to do it in the same way given their volume and variety because data are now much more than simple traceable operations. Data are now considered the main element of competitive advantage so that, in order to fully exploit them, the analysts are learning to implement Big Data strategies and techniques in every area of business. The management of Big Data serves, in fact, not only (as in the past) to record the company's market operations, but also to maintain relationships with customers, to maximize the benefits that a customer can bring to the company and vice versa, as well as to improve the performances in all areas of the company.

An example of this is represented by the so-called *learning from the past algorithms*,[17] such as those used by Amazon, to provide advice to its users based on what other users have researched or purchased, or by Google to recommend which website to visit, or by LinkedIn or Facebook, to suggest to users with whom to connect based on communalities, or relationships with other members of the network, similarities in the profile, in the experiences, in the location or others. According to McKinsey "Big Data can become a new type of corporate asset that will pass through business units and functions representing the basis for competition".[18] Companies today exploit Big Data to obtain information. For example, data on comments, reviews and user preferences posted on social media are used to better predict consumer decisions. The banks collect and study customer transactions to target their advertising campaigns or for credit scoring and fraud detections.

Indeed, many companies in recent years have invested billions of dollars in Big Data, financing the development of software for data management and analysis. This happened because the stronger economies are very motivated to analyse huge amounts of data: currently, there are over 4.8 billion active mobile phones and almost 8 billion people of Internet users.[19] Using Big Data, managers can now accurately measure phenomena and, in this way, can know much more about their business and they can directly translate that knowledge into better decisions and better performances. This new awareness of the importance of a data-based approach also in the business-economic field is well summarized in the formula expressed by the economists Druker and wrongly attributed also to Deming[20]:

> You can't manage what you don't measure.

As already observed, data have no value in themselves, but only if combined with other data and if read through appropriate interpretative models. To create value from data, indeed, it is necessary to implement advanced analytical models, train managers and employees to read analytical outputs, by equipping them with advanced technological tools that are able to transform analysis outputs into business

[17]See Finger (2014).

[18]McKinsey (2011).

[19]https://www.internetworldstats.com/stats.htm.

[20]https://blog.deming.org/2015/08/myth-if-you-cant-measure-it-you-cant-manage-it/.

decisions. Many companies are moving in this direction, launching strategic plans to integrate the various elements and thus focusing on critical decisions, on trade-offs related to different opportunities and on their strategic priorities. These plans will provide support to managers and technicians in order to identify where the greatest profits might come from.[21]

Following these increased needs for data analysis, a new professional figure emerged in these years, called *data scientist*, which subsumes in itself the ability to analyse the data and to extract the value from them, a skill which is typical of the statistician, with the ability to collect and manage datasets which are typical of the IT expert.

Google's chief economist, Hal Varian, said in 2009[22]:

> *The sexy job in the next 10 years will be the statistician. People believe that I'm joking, but who would have foreseen that computer engineers would have been the sexy work of the 1990s?*

In these years, the so-called *data scientists* are developing new analytical and visualization techniques for the treatment of vast amounts of data which, together with the ability to plan statistical experiments with Big Data, will allow in future to fill the current data gaps (e.g. those highlighted in section I.4) and to identify cause-effect links between the various empirical phenomena.

As a matter of fact, the availability of Big Data is currently producing in the world of business what has been called a *managerial revolution.* In 2012, the prestigious Harvard Business Review devoted a special issue to this phenomenon.[23] In the various contributions that appeared in the journal, it was described how once very expensive data-intensive approaches quickly became extremely cheap, making a data-based approach possible for any type of company. The journal also reported the results of a survey (carried out by a joint team of researchers from the MIT Centre for Digital Business of Boston, McKinsey and the Wharton School of the University of Pennsylvania), on the companies that base their decisions on data, in order to test the business benefits. The results showed better performances of these companies compared to traditional firms in all economic sectors, demonstrating that the phenomenon represents a radical transformation of the entire economy and not of just some sectors: no sphere of business will remain exempt from it.

The journal closed with a sentence that should make the veins in the wrists of any manager shake:

> *The evidence is clear: data-driven decisions tend to be better decisions. Business leaders either embrace this novelty or they will be replaced by others who will.*

It is the principle of evolution. In the past, faced with radical ecological upheavals, some species disappeared from the face of the earth and make room for others stronger or simply more suitable to the environment. The manager who does not know how to extract value from Big Data will have to give way to its most advanced version.

[21] McKinsey (2013).

[22] https://flowingdata.com/2009/02/25/googles-chief-economist-hal-varian-on-statistics-and-data/.

[23] Mcafee e Brynjolfsson (2012).

2.7 Some Definitions of Big Data

In the next chapter, I will draw the reader's attention to the methodological aspects related to the Big Data analysis which we mentioned in the previous sections. Before moving to the next chapter, however, I wish to conclude this first chapter by reporting some of the attempts to define Big Data which have enriched the literature in recent years. In fact, it is certainly true that it is impossible to fully define a new phenomenon during its first outbreak and that we have to wait for the dust raised to settle for it to show us the fundamental aspects of the new world that is being created. However, in our continuous search for simplifying paradigms, some of these attempts found in the literature could be helpful to the reader to fix the ideas on some particular aspects treated so far and to better understand them.

In one of the first books that appeared on the subject entitled "*A revolution that will transform the way we live, work and think: Big Data*", the authors Schoenberger and Cukier define operationally Big Data as:

> *... things that one can do on a small scale and not on a large scale to extract new indications or create new forms of value in such a way as to change markets, organizations, relations between citizens and governments and much more.*[24]

Other definitions, also very operational, are those that emphasize the aspects related to the volume. For example, the one proposed by Teradata[25]:

> *A Big Data system exceeds the commonly used hardware and software systems for capturing, managing and processing data in a reasonable amount of time for a massive population of users.*

and the one proposed by McKinsey.[26]

> *A Big Data system refers to datasets whose volume is so large that it exceeds the capacity of relational database systems to capture, store, manage and analyse.*

A little wider is the definition provided by T. H. Davenport referring to both volume and variety:

> *Big data refers to data that is too large to enter a single server and too unstructured to enter a table of rows and columns.*[27]

and the one we find on Wikipedia, which refers to the 3 fundamental V's discussed earlier in this chapter, which reads as follows:

> *A collection of data so extensive in terms of volume, speed and variety that specific technologies and analytical methods are required for the extraction of value.*

---

[24]Schoenberger e Cukier (2013).

[25]https://www.teradata.com/.

[26]McKinsey (2011).

[27]Davenport (2014).

The definition that appeared in *Ively Business Journal*, in contrast, emphasizes in particular the ability to create value, which is connatural to Big Data which we treated in Sect. 2.6:

> *Big data is a company's ability to extract value from large volumes of data.*

In another textbook which is also frequently quoted in the specialized literature, entitled "*Taming the Big Data Tsunami*", we find a definition proposed by Bill Franks who states that:

> *Big Data is an information asset characterized by speed, volume and high variety that requires innovative forms of analysis and management aimed at obtaining insights into decision-making processes.*[28]

In conclusion, however, there is no doubt that the most colourful, but also in some ways the most precise, definition of Big Data that has appeared in recent years is the one posted on Facebook in the year 2013 by the New York behavioural economist Dan Ariely, who reads as follows:

> *Big Data is like adolescent sex: everyone talks about it, nobody really knows how to do it, everyone thinks that others are doing it, and therefore everyone claims to do it.*

---

[28]Franks, B. (2012).

Chapter 3
Statistics and Empirical Knowledge

3.1 The Statistical-Quantitative Method: Measuring to Know

"Without Big Data you are like the blind and deaf in the middle of a highway", asserted the American management consultant Geoffrey Moore.[1] However, we saw in the previous chapter that we can have the availability of an infinite quantity of data, even the huge library of Babel envisaged by Jorge Luis Borges containing all possible books, but without someone to guide us, without someone who takes us hand by the hand, without a "method" of research, on Moore's highway, while seeing and feeling ourselves very well, we run the risk of perishing overwhelmed by the enormous mass of data.

This guide on the data highway has been provided, for centuries, by the statistical-quantitative method to which this second chapter is entirely devoted.

Statistics has always been primarily concerned with defining empirical phenomena and providing them a quantitative measurement so as to reach, if not a full knowledge, at least a summary description of them and an operational approximation. Faced with the complexity of the universe, humankind, soon recognizing his inability to understand it in its entirety, has always tried to circumscribe the phenomena, defining them (tracing precise boundaries around them[2]) and measuring them, thus succeeding in providing at least an operationally useful description. This procedure has a history which is much longer than the history of statistics. We have to mention here again, for the second time in this book, the ancient Egyptian because the pharaohs, in order to support a centralized and bureaucratic state organization, had the need to measure the population and the agricultural production of the vast empire

[1] Moore (2014).

[2] The word *de-fine*, indeed, derives from the latin word *fines*, that means *borders*.

The original version of this chapter was revised: Two figures have been corrected. The correction to this chapter is available at https://doi.org/10.1007/978-3-030-73030-7_6

G. Arbia, *Statistics, New Empiricism and Society in the Era of Big Data*, SpringerBriefs in Statistics, https://doi.org/10.1007/978-3-030-73030-7_3

through a statistical system of censuses. These censuses were conducted every two years under the second dynasty (around 2,900 BC) and then even annually starting from the sixth dynasty (around 2,100 BC). The use of statistics is widely documented also in the Roman Empire. It is the emperor Caesar Octavian himself who informs us about it:

> *... during the sixth consulate I made a census of the population ... In this census four million sixty-three thousand Roman citizens were registered. Then I made a second census with consular power ... and in this census four million two hundred and thirty thousand Roman citizens were registered. And I made a third census with consular power ... in this census four million nine hundred and thirty-seven thousand Roman citizens were registered.*[3]

This system of statistical censuses of the Roman Empire is also documented in the Gospel of Luke:

> *In those days a decree of Caesar Augustus ordered that the census of the whole earth be made. This first census was made when Quirinius was governor of Syria. Everybody went to registered, each one in his own city. Also Joseph, who was of the house and family of David, from the city of Nazareth and from Galilee, went up to Judea to the city of David, called Bethlehem, to be registered together with Mary, his wife, who was pregnant.*[4]

The censuses of the Romans continued even after Augustus with a ten-year frequency, just as we still do today in many countries.

Indeed, the need to measure things to simplify our understanding which is so evidently present in the ancient Egypt and ancient Rome is a characteristic connatural to the human soul. In this regard, I will always remember a funny story concerning a dear cousin of mine, a little younger than me, who, at the age of eight, received a tape ruler as a present. That gift so passionately fascinated the little boy that from that moment on he spent whole days measuring all the objects he came across: the length of a table, the diameter of a plate, the width of a book, the length of my shoe, my nose, my thumb. For a child of that age, the world generates insecurity and appears disproportionate, out of any possibility of understanding and controlling. So, the mere fact of being able to have an instrument capable of providing a quantitative measurement, represented for him an element of safety or, at least, the illusion of being able to simplify a reality that appeared so elusive to him, and the ability to reduce it to some of its quantitative characteristics. That cousin of mine did not become a mathematician nor did he lose his mind in measuring every object with that metre as some of the readers might mistakenly think. On the contrary, he became an established lawyer confirming that the quantitative approach is not for the few initiated individuals with an insane passion for numbers. Each of us, whatever his or her position or profession, is able to appreciate its simplifying power at some point and in some way.

Such an approach has always been present in the history of individuals as well as in the history of humanity. However, it is only with the birth of national states that a quantitative approach in support of the nation management was systematically

---

[3] From the book: *Res gestae divi Augusti.*

[4] *Luke's Gospel 2,1–5.*

affirmed. In this respect, we understand the etymology of the term *statistics* as it is commonly used today. The term, indeed, derives from the term *status*, meaning *political state*. Indeed, the modern conception of Statistics, as a tool for measuring phenomena supporting states' managements and decisions, is largely due to the work of an English economist, William Petty, who in the seventeenth century defined it as "*the art of reasoning through numbers on things concerning the government*".[5]

However, even if measuring and describing are important aspects that defined statistics in its first year, its purpose is not only limited to that. There is a second aim which is much more ambitious. It refers to the possibility of observing reality with the aim of identifying causal links between phenomena, regularities and hidden laws that could be operationally useful in decision-making processes, which can remain hidden at a superficial observation and which can emerge, instead, if subjected to a rigorous examination of the empirical data.

The next section will be dedicated to describing this second purpose.

3.2 The Two Approaches to Reach Knowledge: Induction and Deduction

As stated in the introduction, the purpose of this essay is to present Big Data in simplified terms even to non-specialists in the field. My philosopher colleagues will therefore forgive me if in this section I oversimplify the discussion by affirming that in dealing with the problem of the limits and validity of human knowledge (the so-called *gnoseological problem*), we can proceed by following the two alternative procedures suggested by Aristotle. A first procedure is called *deduction*, a term which derives from the Latin words *de ducere*[6] and consists in fixing some assumptions that need not be proved, relating to the phenomenon that we want to study, and in deriving from them how the observational reality should appear. The second approach proceeds in the opposite direction. Starting from the observation of empirical phenomena, which is necessarily limited, a generalization is attempted reaching statements that can possess, as much as possible, a universal value. This second procedure is called "*induction*" (from the Latin *in ducere*).

As Aristotle defines it, induction is:

> *the process that leads from the particular to the universal.*[7]

In place of the term "universal" or "general" (i.e. what is universally valid), philosophers also use the Greek word *noumenon* from the present participle of the verb νοέω, which means *to think*, "what can be thought". In opposition to the term

---

[5] Petty (1690).

[6] More precisely the latin verb *ducere* means "to conduct" so deduction means "conduct from the top".

[7] Aristotel (1989).

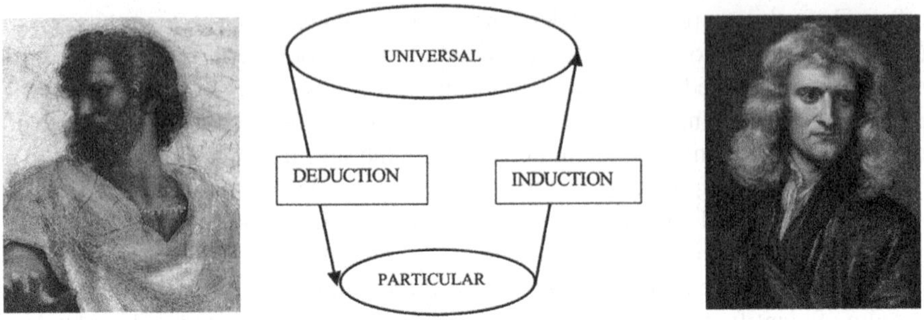

Fig. 3.1 Two basic procedures to reach knowledge: Deduction and Induction

noumenon, philosophers use the term *phenomenon* from *fainòmenon*, participle of the verb *fàinomai*, which means *to appear*, "what appears to our senses".[8]

The two fundamental procedures of reaching knowledge are summarized graphically in Fig. 3.1 together with two emblematic figures representing the two ways of reasoning: Aristotle and Newton.

In Fig. 3.1, Aristotle, the father of formal logic, represents symbolically the deductive way of thinking. Indeed, while admitting that knowledge can be pursued, in principle, following both procedures (starting from a knowledge of the particular to go back to the universal, or starting from the universal to go to the particular), Aristotle believes that a really grounded knowledge can only come from the second approach, provided it makes its moves from true premises.

In the particular there is no science.[9]

However, since the premises of a deductive reasoning, as a rule, are unprovable, he suggests that they must be derived from the intuition of the intellect, thus requiring no proof.

An example of deductive reasoning is represented by the syllogism (also called *Aristotelian syllogism*), a logical procedure based on which, starting from two premises, a conclusion can be derived.

A typical example of syllogism is the following:

- major premise: all men are mortal;
- minor premise: Socrates is a man;
- conclusion: Socrates will die.

The second character displayed in Fig. 3.1 (and taken as an emblem of inductive reasoning) is Sir Isaac Newton. Indeed, while the Aristotelian canons remain prevalent for centuries, they were radically challenged starting from the seventeenth century, by philosophers and scientists (such as Francis Bacon and, before

[8] The concept of *noumenon* is, indeed, also at the basis of the metaphysics in Plato. It is also present in Kant's philosophy, which refers to it as a *hidden reality that lies behind the phenomena we observe beyond their, sometimes deceptive, appearances.*

[9] Aristotel (1933).

him, Galileo) who, moving from idea of the experimental method (also known as, precisely, the *Galilean method*), reconsidered the role of induction and empirical observation in scientific research. Everyone knows the anecdote of Newton and the apple popularized in a colourful way by Voltaire in his *Lettres philosophiques.*[10] Reading the story told by Voltaire, we imagine the scientist sitting under an apple tree in his country estate in Woolsthorpe-by-Colsterworth, resting from his studies, when an apple hits him on the head (legend says that it was the *annus mirabilis 1666*). While everyone else would have limited himself to just eating the apple, the scientist wonders why the fruit has followed that particular movement and speed when falling down, why an apple always falls perpendicularly to the ground towards the centre of the Earth and never to the side or to the high or in any other direction, and why other bodies, such as the moon and stars, are not subject to the same fate. Speculating deductively in that particular case would not have provided him with an answer. However, given that it would have been too long to wait for other apples to fall down (or, better, to wait for a *large number* of apples to fall down so as to be able to generalize the process and reach some conclusions), we imagine the scientist closed in his Cambridge laboratory by sliding a ball on an inclined plane a large number of times (keeping under control the weight of the ball, the inclination of the plane, its height together with all other possible disturbing factors), in order to identify how much was typical of the single experiment and how much, instead, was obeying to a more general rule. In that situation, the famous scientist was using the inductive-experimental method. From the repeated observation of the phenomenon (together with his previous theoretical knowledge of physics) with the experimental conditions kept strictly under control, he reached the formulation not only of the *law of attraction of the apple to the lawn*, but of something of an incomparably more general scope, which would have revolutionized classical mechanics, such as the *law of universal gravitation* that governs the attraction of all bodies in the universe stating that it depends directly on the mass of the two bodies (in this case the earth and the apple) and inversely on the squared Euclidean distance between them.[11]

It should be noted that both the examples given before do not represent cases of pure deduction or pure induction. Somehow Aristotle, when stating the major premise of the syllogism that "*all men are mortal*", was using the experimental observation that does not provide (at least so far) any statistical evidence of immortal men. In the same way, Newton, closed in his laboratory, does not entrust his own conclusions only to the basis of the results of his scientific experiments, but also making use of the prior knowledge and of the intuitions of his own intellect.

Newton himself clearly acknowledges the role of previous knowledge when he states: "*If I have seen further, it is by standing on the shoulders of Giants*".[12]

---

[10] Voltaire, J. J. (1734).

[11] Newton (1687).

[12] *Newton, I. (1675) "Letter from Sir Isaac Newton to Robert Hooke". Historical Society of Pennsylvania. Although often attributed to Newton, the sentence has a longer history and it can be traced back* to the twelfth century as it is attributed to Bernard of Chartres: a French Neo-Platonist philosopher. However, according to Umberto Eco the sentence even dates to a previous period, in

The role of prior knowledge joint with empirical observations in the process of inductive inference is also acknowledged by Immanuel Kant when he says in the preface to his celebrated <Critique of the pure reason> : "*Now, these sciences, if they can be termed rational at all, must contain elements of a priori cognition*" (Kant, 1787).

When humankind started developing more and more sophisticated measuring instruments, together with the possibility of collecting the results and justifying them theoretically, the inductive-experimental way of thinking took on in the course of history, first of all in the so-called experimental sciences (such as physics or chemistry) and then, progressively, in all other branches of human knowledge.

The peculiar characteristics of the quantitative method clearly emerge in the inaugural lecture to the first academic year of the twentieth century in the Royal University of Rome (now University La Sapienza). On that occasion, the famous Italian physicist Vito Volterra speaks for the first time about the "*attempts*" that could be made to extend the quantitative method to biology and the social sciences. Here is how he expresses this concept:

> «*The study of the laws with which the various entities vary, idealizing them, taking away from them certain properties or attributing them some others, and establishing one or more elementary hypotheses that regulate their simultaneous and complex variation: this defines the moment when are thrown the foundations on which the whole analytical building can be built ... It is in that moment that one can see the power of the methods that mathematics puts widely at the disposal of those who know how to use them ... Therefore: shape the concepts in order to introduce a measure; then measure; then deduce laws going back from them to hypotheses; deduce from these hypotheses, through analysis, a science of entities that are ideal, but also strictly logical. Then compare them with reality. Then reject or transform the fundamental hypotheses that have already served when contradictions arise between the results of the calculation and the real world; and thus reach new facts and analogies; or from the present state being able to argue what was the past and what the future will be. This is how, in the shortest possible terms, it can be summarized the birth and the evolution of a science with mathematical characteristics*».[13]

We are at the beginning of the '900, a century in which human knowledge rapidly evolves from the "attempt" of which Volterra is referring to, towards a massive use of quantitative-statistical methods in fields such as economics, sociology and many others.

What we are observing in recent years is a further phase of expansion of this inductivist development which has rapidly extended from the experimental sciences to the social sciences and which is now (under the pressure of Big Data) further expanding also to managerial sciences, to human sciences and to every other field of knowledge including that of which we use every day in our individual choices.

This is the great Big Data revolution!

---

particular to the Latin grammarian Priscianus Caesariensis (AD 500), known for being the author of the most famous book of Latin language during the Middle Ages.

[13] Volterra (1920). Some essays are translated in english and can be found in Giorgio Israel, Essays on the history of mathematical biology. E-book.

The empirical-inductivist way of reasoning does not concern any more only the scientist, but also all of us in every field of social and individual action. If we want to take full advantage of it, we then must understand its rules.

The use of the quantitative method outside of the experimental sciences, however, encounters an objective non-negligible obstacle in the impossibility of carrying out experiments (in fields such as economics or sociology), in a similar fashion to what happens in physics or in chemistry.[14]

We emphasize this important aspect of method using the "lightness" of the words of Milan Kundera:

> «Any student in physics can test the accuracy of a scientific hypothesis. The man, on the other hand, living only one life, has no chance of verifying a hypothesis through an experiment, and therefore [... Tomáš ...] will never know if he should or should not listen to his own feeling[15]».

The character of the novel by Kundera, Tomáš, had to decide whether to leave Zürich, where he moved, to return to his hometown (Prague, which in the meantime had been occupied by the Soviet army) or to remain in the quieter and safer Zürich. However, Kundera tells us that, in order to take such a decision, he couldn't run an experiment, to verify which of the two is the right choice.

In the social and human sciences (in economics, sociology, psychology, law), as is the case with the Kunderian Tomáš, there is usually no possibility of repeating experiments, and, on the contrary, we have almost invariably just a single observation of the phenomenon which can never be repeated in the same experimental conditions ("*ceteris paribus*" as the Latin said) as a truly scientific approach would require.

Let's think, for example, of the use of Big Data in a field which is very far from the experimental situation, such as that measuring a phenomenon like epidemic diffusion.

It is clear that in this case, despite the enormous quantity of data deriving from official health registers, crowdsourced information, data from the web and other sources, it is not possible to repeat the observations in experimental conditions so as to identify with absolute certainty what will be the development of the disease or to measure scientifically the effectiveness of different health policies.

Here, it emerges the second, most ambitious, use of statistics: designing a method to extract knowledge about all reality having access only to a part of the information.

3.3 Obtaining Knowledge from a Partial Observation: *Good* Samples vs. *Bad* Samples

As it is impossible to read all books in the Babel library, given that in practice we can never observe the whole reality (in all its past, present and future states of

[14] Although this limit has been partially overcome in recent years with the advent of the so-called experimental economy. See in this regard the work of the 2002 Nobel Prize winner Vernon L. Smith (2008).

[15] Kundera (2004).

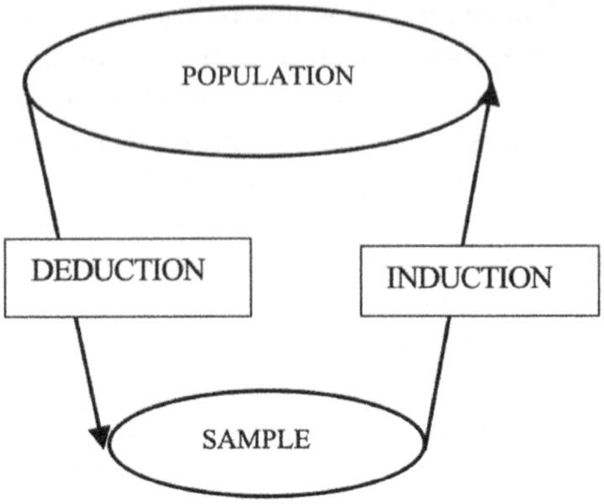

Fig. 3.2 Deduction and induction in the statistical terminology

nature), we are forced to observe only a part of it, and we have to try to draw general considerations from a smaller subset of data. Under this respect, our Fig. 3.1 modifies into Fig. 3.2 where we change the language, and we shift to a terminology which is more familiar to statisticians. The universal (what we would like to know) is called by statisticians the «population», the entire set of data, while the particular (what we are practically able to observe empirically) is what is called the «sample». Therefore, we make a sample observation when, instead of observing all the units of a population, we consider only a subset of them. The statistical induction, therefore, consists in the challenge to generalize what we observe in a sample to the entire population the sample is drawn from. It is clear that if we could observe many repetitions of the same phenomenon (many samples), as it happens in experimental situations such as the repeated observations of Newton's ball on the inclined plane, we would only have to repeat the experiment a large number of times so as to reach satisfactory conclusions. But if this is not possible and we have, in contrast, only a single sample available, we must proceed in a different way.

The problem of extending the inductive approach from the experimental to the non-experimental sciences was solved in the 1930s by Sir Ronald Fisher (among many others), rightly considered the father of statistical science as we conceive it nowadays. In his book (celebrated as a sacred text by statisticians) «*The design of the experiments*»,[16] Sir Fisher is among the first to affirm that, for a statistical experiment to lead to a satisfactory generalization, the sample experiments must be rigorously programmed.

Without going too much into technical details that go far beyond the scope of the present discussion and for which we refer to the specialistic literature,[17] there are some specific characteristics that statistical induction must satisfy.

The most important of these is that the observations cannot be collected chaotically *as they come* and must, in contrast, be selected according to a predefined "design" which obeys some precise roles. One important example is when the data collection

[16] Fisher (1935).

[17] E. g. Azzalini (1996).

follows a sampling system called *random* in which, by definition, each unit has the same probability of being drawn,[18] thus guaranteeing, so to speak, the maximum of objectivity in the data collection criterion.

If we think carefully about it, the claim of being able to predict what happens in a whole set of situations observing only a subset of them, looks like pure magic (or craziness or boasting according to our attitude). Imagine, for instance, that you are having a dinner with friends and you take a picture of the group of people gathered around the table. Then, I call you on the phone and I ask you to send me only a small portion of that picture (e.g. the pixels in the bottom left corner of the image) and I claim that I will be able to say who else is sitting around the table! You would most probably think that I am a deceiver or a boaster.

But this is exactly what happens with sampling!

The crucial role in guaranteeing reliable results is played by the criterion we follow when we draw the sample units. This criterion is what we call a *sampling design.* The sample design called *random,* as we said before, is the one that guarantees that all individuals in the population have the same chance of being drawn, and thus, it is the criterion that minimizes the possible distortions that could derive from a partial observation of the phenomenon. It is impossible, and in any case outside the range of interest of the present essay, to describe properly in few words how to properly design a rigorous sample selection: this is the job of a survey statistician. However, it is conversely extremely easy, and understandable for everyone, to explain what are the characteristics of samples that are *badly* designed and why in those cases statistical inference is doomed to fail.

Let me try to achieve this aim presenting a series of examples.

The first is related to electoral forecasts. Electoral forecasts attract a large public interest and are, indeed, a very good example to evaluate the appropriateness of statistical inductive inference. In fact, in contrast to many other sampling situations, everybody is interested in having anticipations about the electoral results in the weeks before the elections, but everybody can also easily verify after the electoral counting if the forecast was correct or not.

One of the first experiences of electoral forecasts dates to the work of George Gallup, a practitioner statistician specialized in survey sampling for measuring public opinion in the United States in the first half of the twentieth century. His name is linked to the great success he obtained when he correctly forecasted the results of presidential elections with sample data, especially when, in 1936, he managed to predict the success of the Democratic candidate Franklin Delano Roosevelt against the Republican Alf Landon using a sample of only 50 thousand electors. This success gave him a particular popularity because the Literary Digest (a widely influential American magazine that up to that point was considered the most reliable in terms of electoral forecasts) failed the prediction using a much larger sample. Indeed, Gallup was even able to predict the amount of error made by the Literary Digest.

---

[18] In general, a sample design does not necessarily require that each individual has the same probability, but it is enough that we are able to exactly calculate the probability of being drawn for each individual of the population.

But his name is also linked to a big failure.

In fact, twelve years later, in 1948, Gallup failed dramatically the prediction at the new presidential elections. At the time, Harry S. Truman, the Democratic candidate, had ascended to the presidency in April 1945 replacing Franklin D. Roosevelt after his death and there were big expectations on the elections because they were the first after the end of the Second World War. The opponent of Truman was the republican candidate Thomas E. Dewey. This anecdote is very popular among statisticians because one of the latest reports of Gallup foresaw 46.5% for Dewey to 38% for Truman. The statistician was so confident of his predictions that he communicated them to the press and the *Chicago Daily Tribune* (together with several other newspapers), before any polls closed, published the results in order to anticipate the election of the new president of the United States to the whole American population and to the world. A celebrated representation of this episode is the picture reported in Fig. 3.3.

Despite the fact that the newspaper printed as its headline "Dewey defeats Truman" crediting Gallup's forecasts, the man in the picture was indeed Harry Truman and his satisfied expression is more than justified by the fact that the forecasts were wrong. Gallup believed the error was mostly due to the fact that his sample observations

Fig. 3.3 1948 American presidential elections. The grinning happy man in the picture is Harry S. Truman. The president elect holds up a copy of the newspaper that wrongly announces his defeat

ended three weeks before elections and he missed a number of electors that changed their minds in the last two weeks, but there was much more in it.

What happened was that Gallup's predictions were mostly based on a sample of individuals that were contacted through the telephone. Even if it is difficult to believe now especially for the younger people, at the time only a few people had a private telephone at home[19] and they were, not surprisingly, the wealthier individuals that could afford an instrument that at the time was still quite expansive. Therefore, Gallup was not observing an objective picture of reality. By cancelling a priori the possibility of observing people without a telephone, he was over-sampling wealthier people and, in contrast, under-sampling poorer individuals. He was thus violating the condition of equal chance of being drawn which is, as we said before, one important characteristic required by statistical theory to have reliable sample observations. Wealthier people at the time were more incline to vote for T. E. Dewey, and for this reason, Gallup was sampling mostly from the population of Dewey's supporters and the result was then obvious. Collecting data through the telephone was a convenient way of gathering quickly sample information, but it was not a rigorous statistical sample and, as such, was not useful for drawing theoretically grounded inductive inference on the future electoral results.

Similar errors were repeated more recently by American pollsters. In the 2016 presidential elections, most of the pollsters were predicting a large success of Hillary D. Clinton, a prediction that turns out to be wrong with the final success of Donald J. Trump. This was considered by someone as a big failure of statistics that shed a negative light on the actual possibility of using samples in order to anticipate electoral results.

What happened in that case was that predictions were based on samples drawn from a list of those that were classified as the *likely voters*. Such a list could be compiled long before the elections because a likely voter is defined as someone who decided to take part in the last three presidential elections, that is, in the last 12 years. However, in 2016, there was a big change in the attitude of potential electors with over 7 million voters more than those in 2012. On that occasion, the percentage of voting-eligible population increased from 54.9 to 59.2%. Furthermore, many of those that didn't decide to vote in the past were attracted by the figure of Donald Trump and decided to participate in the 2016 election. In contrast, many electors that had been assiduous voters in the past (mainly democrats) decided to desert the polls. The result was that most sample surveys weren't able to capture a large part of Donald Trump's electors thus dramatically underestimating his votes and wrongly predicting Hillary Clinton's success.

In the most recent 2020 presidential elections, most of the pollsters changed their strategy and relied, more cautiously, on the notion of *registered voters* as the basis for drawing a sample. Registered voters are defined as those that in the last weeks before the elections register themselves, thus explicitly manifesting their willingness to participate. This method was less convenient than the use of likely voters because

---

[19] It is calculated that, by the end of the Second World War in 1945, there were five people for every working phone (https://phys.org/news/2019-03-fall-landline-years-accessible-smart.html).

it involved waiting when most of the people are registered, but, as a consequence of a better statistical sample design, this time most of the surveys successfully predicted Joe R. Biden's success.

A second example of the failure of statistical induction when the sample is not rigorously programmed is the recent Sar-Cov-2 pandemic. During the healthcare emergency caused by the worldwide diffusion of the coronavirus started in late 2019, everybody was interested to have timely estimates of a series of epidemic parameters such as the percentage of infected people and the fatality and mortality rates among the whole population.[20] Although the worldwide community of statisticians repeatedly warned against the improper use of the available data to make sound inductive inference, these parameters were customarily calculated on the basis of the information derived from the medical swabs that were administered by the local health authorities in all countries on an emergency basis. However, also in this case (like in the previous US election examples), the set of data does not come from a rigorously programmed sample, but only from a collection criterion which was dictated by pure convenience. Obviously, the medical swabs were administered (at least in the first phase of the explosion of the pandemic) mostly to people displaying severe symptoms and not to people without or with few symptoms. As a result, the data were inappropriate to correctly predict the true fatality and mortality rates: the datasets were helpful only in estimating the number of people with symptoms and not that of the total infected people. As a consequence, the proportion of infected people was obviously higher in the population than in the sample so that both the fatality and mortality rate of the virus were overestimated. On the other hand, the underestimation of the number of infected people had severe consequences in terms of the pandemic health surveillance with a much higher diffusion of the virus than that emerging from official data and with a very large number of undetected asymptomatic infected people who represented a vehicle of the virus diffusion. In fact, many studies, using different methodologies, estimated that the number of infected people was much larger than the one estimated with the official datasets based on the medical swabs.[21]

The examples reported before clarify that we should therefore avoid drawing our conclusions based on samples obtained following various criteria dictated by convenience and ease of collection and not by statistical theory. This condition, however, is seldom satisfied in the case of Big Data, when data are usually unstructured (see

[20] The fatality rate is defined as the percentage of deaths among infected people while the mortality rate is defined as the percentage of deaths over the whole population.

[21] See, for instance, the findings of the "Center for Disease Control and Prevention" of New York (https://www.nytimes.com/2020/03/31/health/coronavirus-asymptomatic-transmission.html), of the Chinese government (https://www.dailymail.co.uk/news/article-8140551/A-coronavirus-cases-silent-carriers-classified-Chinese-data-suggests.html), of WHO https://www.who.int/docs/default-source/coronaviruse/situation-reports/20200306-sitrep-46-covid-19.pdf?sfvrsn=96b04adf_2, of the World Economic Forum https://www.weforum.org/agenda/2020/03/people-with-mild-or-no-symptoms-could-be-spreading-covid-19/ and the scientific papers published by Li et al. (2020) and Bassi, Arbia and Falorsi (2020).

Chapter 2) and collected *as they come* without any precise plan of experiments only following criteria of pure convenience.

Indeed, many of the new, non-traditional Big Data sources suffer from this limitation. For instance, a popular way of gathering large quantities of data is through the already quoted operation called web scraping (or also web harvesting) that is the operation of extracting data from websites and reshaping them into a structured dataset. The technique consists in writing software codes and using a process (called *robot* or simply *bot*) to access the Internet directly through a web browser in order to automatically extract the data from a large number of websites. While this is a very good way in which large quantities of data can be quickly accumulated in an easy format already available for any subsequent analysis, this is rarely a good way of building up well-justified statistical samples.

For instance, if we are interested in gathering information about house market to have an idea of average prices in one specific area, we can think of harvesting the web and visit all the real estate websites to download the relevant data. This is certainly an easy way of quickly accumulating a lot of information, but the dataset obtained in this way cannot be considered a proper statistical sample as we described before, and we run the risk of wrongly estimating the average price for a number of reasons. First of all, in our sample, we lose the more expansive houses and all those houses that do not enter the market and are bargained directly by the seller and the buyer. Secondly, we run the risk of counting more than once the same apartment when this is advertised by several real estate agencies. Finally, we may overestimate the average price because we consider the supply price as it appears on the advertising websites and not the actual final price which is likely to be much smaller after the negotiation between supply and demand.

A similar situation occurs when the Big Data are gathered through the operation called *crowdsourcing* (a term which we have already introduced in Sect. 2.3) which refers to data voluntarily collected by individuals. An example is represented by data collected through smartphones, in order to measure phenomena that are otherwise difficult to quantify precisely and timely. In a recent project conducted by the Joint Research Center of the European Commission in Seville,[22] crowdsourcing techniques were adopted to gather timely data on food prices with the aim of anticipating food crises in developing countries where official data are either not enough trustworthy or, in any case, collected without the desired temporal and geographical disaggregation. In particular, the aim of the JRC crowdsourcing exercise was to assess the potential of this form of data collection and to establish a quality methodology to efficiently produce reliable geo-referenced data on food prices at the local and regional level, accessible in real time, in order to meet the data needs of governments, food supply chain actors and other institutions. In particular, the quoted work refers to an initiative for collecting data on food commodity prices via citizen contributions submitted in Nigeria using a mobile *app*: volunteers were required to submit actual transaction prices via their smartphones. Although very useful in order to achieve the said scopes,

[22] Arbia et al. (2020).

these data set could not be assimilated to a proper statistical sample for a number of reasons.

First of all, in order to increase the willingness of the crowd to participate in the initiative, a gamified reward system was included where valid daily submissions were rewarded. This fact, however, induced possible fraudulent behaviour from the participants to the initiative. Indeed, from a statistical perspective, mobile phone numbers are not an ideal sample frame for observing units from a target population in that different links can be established between mobile phone numbers and individuals. To establish a one-to-one relationship, thus and avoiding multiple relationships, the initiative allowed a phone number to be registered only once. However, it was inevitable that a person with several phones could send data from all of them to enjoy more rewards. For the same reasons, the survey run the risk of possible fraudulent activities of collectors either in the form of non-independence (collectors that share the same picture to be sent) or in the form of duplications (collectors that send more than once the same picture in order to obtain more rewards). Errors may also derive from different sources such as wrong interpretations of the collectors of the phenomenon to be observed and locational errors due to mistakes in the recording of the coordinates. A further problem derives from the digital divide which might be still a problem in some countries. In fact, even if accessible broadband mobile technology was available in most parts of the country considered in the Nigeria exercise, Internet can provide the infrastructures and solutions for enabling the development of a crowd, but the diversity of opinions may be limited by the technological inequality.

All in all, the experience was judged as positive, but the statisticians involved in the project had to run several preliminary analyses and to introduce several corrections in the dataset before they could be used for monitoring and forecasting purposes.

The lesson that we learn from all the examples reported in this section is that it is not enough to collect data to call them a sample and to use them for a generalization to the whole population. A sample collection has to be rigorously planned from a statistical point of view in a similar fashion of what happens to a scientist who plans her experiments to reach convincing conclusions about general laws.

The idea that empirical observations must be collected following a strict criterion can already be found in the philosophy of Immanuel Kant who states:

"When Galilei experimented with balls of a definite weight on the inclined plane, when Torricelli caused the air to sustain a weight which he had calculated beforehand to be equal to that of a definite column of water, or when Stahl, at a later period, converted metals into lime, and reconverted lime into metal, by the addition and subtraction of certain elements a light broke upon all natural philosophers. They learned that reason only perceives that which it produces after its own design; that it must not be content to follow, as it were, in the leading-strings of nature, but must proceed in advance with principles of judgement according to unvarying laws, and compel nature to reply its questions. For accidental observations, made according to no preconceived plan, cannot be united under a necessary law".[23]

[23] See the Preface to the second edition to the <Critique of the pure reason>, Kant (1787).

Notice that the German philosopher in the eighteenth century uses exactly the same terminology that we use nowadays in statistics, by referring to terms like "design" and "preconceived plan".

The same idea is further clarified in the following passage by the mathematician and philosopher of science Jules Henri Poincarè, taken from his book "Science and method":

> The scientific method consists in observing and experimenting; if the scientist had infinite time, there would be nothing but to say <Look, and look carefully> , but since he lacks the time to look at everything and even less to look carefully - and it is better not to look than to look badly - finds itself in the need to make a choice. Knowing how to make this choice is therefore the first question. This problem arises as much for the physical as for the historian; and it also arises to the mathematician, and the principles which must guide them are not without analogies.[24]

In this passage, there is a notation that strikes me particularly and that I believe perfectly clarifies the relationship between Big Data collected without an experimental design and rigorously planned empirical observations, and, therefore, describes perfectly the period we are living: "It is better not to look than to look badly".

This is an extremely important comment to keep in mind the use of Big Data to direct our choices. Is it not looking badly at things if we rely uncritically on Big Data without questioning their veridicity[25]? Or if we give to a set of empirical data, however collected, the power to guide our choices?

According to Poincarè, therefore, in order to follow a scientifically based approach, it is preferable to have a limited number of data and observations whose quality is under our control ("*knowing how to make the choice*"), rather than entrusting our conclusions and our operational decisions to a mass of data that reaches us uncontrollably. The French philosopher (in a period in which it was impossible to foresee the data deluge situation in which we find ourselves today) warns us about selecting some of the available data, discarding others following a strict criterion, rather than bulimically accumulating them all until they lose track of their actual information content.

In summarizing many of the ideas we have tried to express in this chapter in a synthetic formula, Poincarè also provides us with a valuable indication for not getting lost in the narrow corridors of the immense Babel library, distracted by the many volumes that are absolutely meaningless and that are distributed without any order in its infinite shelves without any cataloguing. If we are looking for a specific book in the hexagons of the library, we have to follow a precise path. This path is indicated by the idea of sampling design which we have described in this chapter.

Before we close this section devoted to the strategy of acquiring knowledge making use of a partial observation of reality, there is a further point that needs to be stressed to remark the difference between a theoretically grounded statistical

---

[24] Poincarè J.-H. (1908).

[25] See Sect. 2.5.

inference and the ineffable art of a magician (or self-styled such) who interprets the messages of a sorcerer stone.

In fact, it is important to keep in mind that our inductive conclusions can never be reached with absolute certainty, but only with a certain level of probability. Such a probability can be very high if we select our sample units in the best possible way, but, in any case, never equal to 1. In fact, we have always to remember that what we have in hand after a sampling exercise is only one of the many possible samples that we could have randomly drawn with a given probability. Therefore, at the end of the induction process conducted with the statistical method in non-experimental conditions, we can only discern between two contrasting working hypotheses the one that is more likely to be true on the basis of the empirical observations.

In the process of induction from samples, Ronald Fisher makes use of a fundamental concept called the *principle of repeated sampling*. Adelchi Azzalini describes this idea in this way: "The term *repeated sampling* derives from the fact that if we replicate the sample extraction for many and many times ... [... we would find ourselves in an experimental situation. *ed. note*]. In the vast majority of real situations this replication of the experiment does not take place, but we reason as if it occurred".[26] Following this procedure, therefore, we will base our conclusions not on the results of our calculation in the single sample which we observed (inevitably subject to the randomness of the sampling operation), but on the random characteristics that this calculation could theoretically assume in a large number of possible experiments. These random characteristics are studied at a theoretical level by mathematical statistics which provides the practitioners with a series of important results as a guidance in choosing the best way of making inductive inference.

The *principle of repeated sampling*, by referring to a theoretical situation that we could define as *pseudo-experimental*, allows us to overcome the obstacle of not having the possibility of repeating the experiments in the social and human sciences like we do in the hard sciences, and thus to extend the idea of inductive knowledge also to these fields as long as we are ready to accept answers that are not certain, but only probable.

3.4 Some Epistemological Ideas: Inductivism, Falsificationism and Post-Positivism

As we argued in the previous section, an approach to knowledge which is based on empirical observation has been widely used for centuries in the exact sciences and has been extended to social sciences dating from the beginning of the last century. The new fact introduced by the Big Data revolution is that this kind of approach has been spreading, in recent years, not only to assist the choices of the single economic operator (who, as we have seen for example in Section I.5, recognizes in this a competitive advantage), but also, increasingly, to those of the individuals, who, in

[26] Azzalini (1996).

their everyday decisions, assume a new attitude in which the empirical data play a fundamental role.

In this sense, anyone, in their choices, has to develop, implicitly or explicitly, some empirically based decision-making strategy.

Indeed, in recent years, we find ourselves, both as individuals and as a society, following a path similar to the one followed by the philosophers of science during the twentieth century, increasingly asserting a sort of neo-empiricism that could be defined as "social-empiricism". Following this new approach, all knowledge seems to be increasingly based on empirical experience and therefore on the ability to collect and elaborate observational data.

Although this is not a philosophical essay, it is worth discussing in this chapter, albeit briefly, some of the fundamental passages in the philosophy of science, which can provide us a guide to that social-neo-empiricism towards which each of us is unconsciously moving in these last years.

It should be noted, to begin with, how philosophical thought in the epistemological field has evolved and changed significantly over the last century. At the beginning of the twentieth century, in fact, on the basis of the contributions of thinkers such as Hume, Comte and Mach, it was established the epistemological school of thought called *neoempiricism* (or *neopositivism* or *logical empiricism* or also *inductivism*) which played a predominant role in the philosophical literature at least until the mid-twentieth century. The followers of this school of thought, born in the so-called *Vienna circle*, affirmed the need to base all knowledge on empirical experience. According to these scholars, starting from empirical observations, we can confirm our working hypotheses and thus we can predict the future development of the various phenomena.

In opposition to the neo-empiricism, the Austrian philosopher Karl Raimund Popper rejects the idea that the scientific hypotheses can be validated only on the basis of the empirical data. Furthermore, by introducing the distinction between *science* and *pseudoscience*, he contrasts the idea of inductivism with that of *falsificationism*. In the nutshell, since we are not able to verify a universal hypothesis (which, by its very nature, is related to a theoretically infinite number of cases observable in future) by using only the relatively small number of data (necessarily finite) observed up to now, Popper claims that it is more correct to subject all theories to a criterion of *falsifiability*. In this sense, we will let (temporarily) survive only those theories that resist many and rigorous attempts of falsification. Popper's idea is that a theory can be definitively rejected (in his terminology: falsified), but never definitively accepted. If a theory resists various attempts of falsification conducted with *audacity* and *severity* by overcoming all empirical controls, then it can be said that it is *corroborated* and, therefore, provisionally admitted in the body of our knowledge.

The distinction between the Popperian approach and the neo-empiricism of the Vienna school, which is fundamental in epistemology, becomes more nuanced when they are used to take decisions in the everyday life. In fact, in this case, it is not necessary to reach universal truths, but only pseudo-scientific truths, which can guide us in the best way to take our everyday decisions on the basis of the empirical data. In this sense, an individual, far from searching for universal laws, should only make use of statistical laws which describe cause-effect links between events with a certain level of probability. It is the Popperian idea of *verisimilitude*, (the German word used by the author), which coincides with the statistical term *likelihood* introduced in the literature by the aforementioned Sir Ronald Fisher.[27] As mentioned in the previous section, taking up an idea which is originally due to the Danish statistician Thorvald Thiele,[28] Sir Fisher suggests the use of a likelihood function which allows to identify, between two contrasting hypotheses, the one that is most plausible on the basis of observational data. According to Fisher, no experimental research can prove the truth of a theoretical hypothesis, but only to state whether it is more or less likely than an alternative hypothesis used to contrast it. The convergence between Fisher's *likelihood principle* and Popper's idea of *corroboration* is also recognized by the latter when he states:

> We can interpret [...] our measure of the degree of corroboration as a generalization of the Fisher likelihood function.[29]

Therefore, starting from different assumptions, for both authors, the "Truth" is only a regulative ideal to which it is certainly necessary to strive, but abandoning the ambitious goal of attaining it through empirical experiences (which are by their very nature only limited) and being just satisfied to obtain the best approximation to it thus accepting the most likely theory based on observational data (in other words the one that is closest to the perfect explanation of the phenomenon), as a practical guide to our decisions.

Along the same lines of thought, we find the contribution of Carl Gustav Hempel, who states that, in the explanation of empirical phenomena (and therefore in making decisions supported by data), we can use the laws determined by what he calls *statistical probabilities*, which are defined as probabilities determined on the basis of the observed frequency. These *statistical probabilities* help us to formulate statistical inductive explanations that are certainly not universally valid, but, in any case, operationally useful. The Hempelian approach to epistemology echoes in statistics the so-called frequentist approach, ascribed to statisticians such as Jerzy Neyman, Egon Pearson and again Ronald Fisher among the many others. The so-called frequentist (or Fisherian) statistical inference is associated with a frequentist interpretation of probability, which in turn defines the probability as the relative frequency of an event in a theoretically infinite number of experimental observations. For example, we may think of the probability of obtaining a "head" when tossing a coin as its relative frequency if we submitted it to an infinite number of trials.

During the last century, the Hempelian model based on statistical probabilities has been subjected to criticisms of various kinds which refer mainly to the fact that very often the events we are interested in are rare and, as such, present a low probability of occurring. Consider, for example, a medical doctor interested in explaining the onset of a disease that has a relatively low frequency in the overall ensemble of

[27] Fisher (1922; 1925).

[28] Thiele (1889).

[29] Popper (1959).

individuals (e.g. lung cancer). He obviously cannot rely on the cases that arise with high statistical probabilities calculated on all individuals but will have to look instead at both the relative frequencies conditioned to the exposure to a certain specific risk (e.g. smoking), and to those of the cases subject to non-exposure to the same risk factor. These probabilities subject to constraints, in the statistical language, are called *conditional probabilities*. In this respect, the philosopher Wesley Salmon proposes a model which is known as the model of *statistical relevance* in which this intuition takes shape. Salmon's model is theoretically grounded on a very well-known statistical approach, known as the *Bayesian approach*. This approach derives from the name of the Reverend Thomas Bayes, a mythical figure in the statistical Olympus, who in the second half of the eighteenth century proved a famous theorem that was published only two years after his death. The Presbyterian monk certainly did not imagine, in elaborating the mentioned theorem, that an entire school of thought would have been originated from it. Similar to Salmon, a follower of Bayes believes that the degree of trust we have in a working hypothesis can be measured by its probability. However, this probability is derived as a complex mechanism where both prior knowledge and empirical observation play a role. In fact, at the beginning of her investigation, a scientist, or anyone who poses a decision problem based on experience, should determine what are the initial probabilities (which a statistician calls *prior probabilities*) of the hypothesis under consideration. It is only in a second moment that the researcher should modify such probabilities in the light of the observational data in order to determine the so-called *final (or posterior) probabilities* on the basis of which the final decision will then be taken. It is clear then that the Bayesian interpretation of probability differs substantially from the frequentist definition given above. For a Bayes follower, the probability expresses only the *degree of belief* we have in the occurrence of an event. This degree of belief is based on the prior knowledge we have about the event (derived, for instance, from previous experiments or our personal opinion on it) modified with what we learn from empirical data. This definition has an advantage and a disadvantage with respect to the frequentist definition. Indeed, a frequentist can only assign a probability to events that are replicable (at least theoretically) an infinite number of times while a Bayesian can always obtain an evaluation of the probability of an event. On the other hand, a frequentist definition of probability is an objective evaluation in that, after an infinite number of experiments, everybody reaches the same conclusions, while a Bayesian probability is more subjective, because the prior knowledge differs from one researcher to the other.

It is interesting to note how the two definitions of probability, which are so important in statistics and lead to two different school of thoughts on how to lead inductive inference, are echoed in the philosophical dualism between Hempel and Salmon in epistemology.

The neo-empiricist conception and its derivations have long remained prevalent in the philosophy of science until the middle of the last century when they were subjected to the criticisms of the so-called post-positivist epistemologies supported, between the 1950s and the 1980s, by scholars such as Thomas Samuel Kuhn, Gaston Bachelard and Bastiaan Cornelis van Fraassen.

Thomas Samuel Kuhn was an American scholar who in the first half of the twentieth century gave important contributions to the history and philosophy of science. He is responsible for introducing an epistemology in sharp contrast to that of Karl Popper's falsificationism, with which he argued for a long time.

His celebrated book *The Structure of Scientific Revolutions* (Kuhn, 1962) was highly influential both in the academy and in a more widespread audience, and had the merit of introducing in the common language the new term of *paradigm shift*.

The quoted book represents an important watershed between the neo-empiricist and post-positivist philosophy. Kuhn, like many other post-positivist philosophers, refers to a fundamental distinction between *mature sciences*, which are characterized by a widely shared paradigm, and *immature sciences* that do not contemplate such a paradigm. For example, physics, biology and chemistry are among the first in reaching the stage of mature sciences, while economics, social sciences and psychology are among the latter. In the mature sciences, the task of scientists is only to complete a system of shared paradigms at least until the emergence of some new revolutionary theory (e.g. in physics, the emergence of the Einsteinian theory of relativity). On the contrary, the immature sciences are characterized by persistent phases in which different theories try to impose themselves on the others never reaching a commonly shared and accepted paradigm. The individuals (both street men and managers), who use Big Data to make their own decisions, behave like a scientist of an immature science where it is not so important to affirm an absolute truth, but only to approximate, better than anyone else, the correct explanations to the empirically observed phenomena.

On the other hand, a modern view admits that also a *mature science* may be subject to a certain degree of uncertainty. Indeed, the Heisenberg uncertainty principle, one of the most surprising and rich philosophical implications of quantum mechanics, underlines the probabilistic nature of a science as mature as physics, replacing the idea of a deterministic physical theory[30] with an indeterministic approach that is certainly only based on valid statistical laws until it is proven otherwise, but that is, however, still operationally relevant when it comes to take a decision. It is here that emerges the idea of the *paradigm shift* which happens when the dominant paradigm, under which science normally operates, becomes incompatible with new empirically observed phenomena, thus necessitating new theories.

Gaston Bachelard was a former postal clerk and then a physicist and a chemist that devoted the last part of his life to philosophical speculation and academic work on epistemology. In his work, Bachelard underlines the possible negative role that can be played by the a priori probabilities by introducing the idea of *epistemological obstacle*. This obstacle is constituted by the set of barriers (linked to ways of thinking, to common sense or to past theories), which prevent us from seeing the truth and

[30] According to the definition of Pierre Simon de Laplace, we call "deterministic" a theory which, on the basis of precise information on the initial state, arrives at determining the state of the system in any future moment of time.

to find explanations for the observed reality. In one of his writings, the philosopher states: "Some knowledge, even correct, blocks too soon many useful research".[31]

We find a very similar critical view, in an even more explicit form, in the reflections of the American theologian Reinhold Niebuhr, who expresses it in this way: "Men rarely learn what they already believe they know".

Starting from this fundamental criticism, Bachelard underlines the importance of what he calls *epistemological rupture* (i.e. the disruption of the senses and the imagination operated by empirical observation) which therefore plays a key role in the development of human knowledge.

In years closer to us, the Dutch-American philosopher Bas Van Fraassen proposes a new epistemological approach, called *constructive empiricism*, which in turn is based on the idea of *empirical adequacy*. According to Van Fraassen (1980), a theory is *empirically adequate* when everything stated by it about observable events is true. Therefore, in order to accept a theory, it is sufficient to believe that it is empirically adequate, in other words that it correctly describes what is observable, regardless of its theoretical adequacy and its comprehensiveness. Consider, for instance, two scientists who place themselves in two opposite theoretical positions: the first one accepts the atomic theory but does not express any opinion about the question of whether atoms exist; the second is absolutely convinced of the existence of atoms. The practical conclusions of the two scientists will be exactly the same because the two positions are empirically equivalent; that is, they are justified by all the observations made up to that point.

Van Fraassen's empirical-constructive model is therefore the one we follow implicitly in making our everyday data-driven decisions which, however, disregard any claim to universality.

In this brief overview of the philosophical thought of science in the last century, we deem it necessary to give a final consideration to the dichotomy between simplicity and complexity. This antinomy has always played a fundamental role in the philosophy of science and in general in the analysis of the scientific method. Indeed, science is often identified as the place of the research of theories which are able to explain empirically observed complexity in a simple and sufficiently adequate way, that is, theories able to bring the complexity of the visible, back to some simple rules that could not be seen at a first superficial examination.

In recent years, however, some theories, known as *theories of complexity*, have been developed in various scientific fields, ranging from mathematics to physics, to biological sciences, to social sciences. Among these, we must certainly mention the *theory of chaotic systems* related to certain deterministic physical systems in which small changes in the initial conditions can cause large-scale big changes. It should be noted that the unpredictability of chaotic systems refers to the complexity of the deterministic laws that govern them and not to that of stochastic physical systems governed by statistical laws, mentioned earlier. The field of economics has also been interested in these aspects, for example with the theory of artificial societies based on the idea that the complex regularities in the behaviour of a social system are the effect

[31] Bachelard (1934).

of few and simple regularities in the interactions between the members of the system. One of the first attempts to give empirical support to these theories in economics is due to the Nobel Prize for Economics Thomas Schelling who, in his volume «Micromotives and macrobehavior»,[32] uses simulation methods to analyse the segregation processes that lead to the formation of neighbourhoods inhabited almost exclusively by individuals of the same ethnic group. A similar idea is contained in the writing, never published, by Durlauf[33] who, by criticizing the prevailing economic school of thought based on the *representative agent's paradigm*, describes an economy as a collection of agents that interact in space and time. In his pioneering work, he suggests a parallel between the economic discrete choice models and the formalism of statistical mechanics (highlighting how the physical models that explain how a collection of atoms display the correlated behaviour which is necessary to produce a magnet) could be used in economics to explain the interactive behaviour of the single agent.

While presenting a great attraction from a theoretical and conceptual point of view for their scientific applications, complexity models are inadequate to support individual empirically based choices and decisions.

It is the simplicity of the synthesis that we should look for in the world of Big Data and in their use in everyday choices: an empirical-constructive scheme that can help us in taking our decisions without drowning in the massive quantity of data to which we are subjected daily.

3.5 The Lost Art of Simplicity: Ockham's Razor and the Role of Statistics

If there is a word that describes well the essence of the statistical method this is the word *simplicity*.

The world is complex by its very nature and we cannot hope to be able to provide an exhaustive explanation in all its many aspects. Moreover, in the era of Big Data, even the way we look at the world, through increasingly accurate tools, becomes more and more complex, requiring new capabilities to manage data and new technical instruments. In this panorama, the important mission entrusted to the statistical method will be more and more that of pre-digesting data through indicators, visualization tools and models so as to make them usable to a growing number of potential users. Reinterpreting the words of Vito Volterra[34] in the light of current challenges, the statistical method will have to answer to the increasingly stringent need to simplify reality through the definition of entities that simplify its interpretation, providing measures (though possibly approximated) and searching for mechanisms that are

[32] Schelling (1978).

[33] Durlauf (1989).

[34] See page 26.

able to describe the nature in a way that is perhaps imperfect, but still operationally useful according to a scheme of *constructive empiricism.*

Indeed, such an approach has been present for centuries in the philosophical thought and it is well summarized in the so-called *Ockham's razor* principle. With such a principle, we refer to the methodological approach, expressed by the monk William of Ockham in the sixteenth century, which suggests the uselessness of formulating hypotheses in a number which is larger than what is strictly needed to explain a given phenomenon. This principle, in its essence, invites us to avoid formulating models that are so much complex, that they run the risk of overshadowing the ultimate purpose of our study which is to explain reality and support taking rational decisions. Therefore, among all the possible theoretically explanations of a phenomenon, it will be appropriate to choose the one that appears the closest to reality without seeking unnecessary complications while trying to reach an unnecessarily more accurate explanation. The razor metaphor has to be understood in the following way: the most complicated hypotheses should be progressively discarded as if they were subjected to a sharp blade. Such a principle of simplicity was already known to all medieval scientific thought, but William of Ockham is credited for having formalized it.

The principle of Ockham's Razor (originally conceived as a purely philosophical concept) was later used also as a practical rule for choosing between competing scientific hypotheses, which had the same ability to explain a natural phenomenon. In fact, given that different scientific theories can produce the same empirical observations, the Ockham razor can be used to discriminate between them. It is also known in Statistics as the *principle of parsimony* that is the principle according to which the model to be preferred for explaining a phenomenon is the one that involves the least number of entities (parameters and hypotheses). In phylogenetics,[35] just to quote an argument which was dear to Ronald Fisher, parsimony is used alongside the principle of maximum likelihood suggesting that the evolutionary tree that best explains the data is the one that requires the least number of modifications.

Indeed, love for simplicity has always been expressed in many different fields apart from those quoted so far in this essay. Leonardo da Vinci, for example, states: "Simplicity is the supreme sophistication", and it is the same concept expressed also by the writer Henry Wadsworth Longfellow: "In the character, in the ways, in the style, in all things, the supreme excellence is simplicity". The German-American expressionist painter Hans Hoffman, having in mind the figurative arts, says that even in that field "The ability to simplify means eliminating the unnecessary so that the necessary can speak". To understand fully the artistic implications of this sentence, just think, for instance, of the meaning of "eliminating the unnecessary" in sculpture.

A few years ago, Josh Holmes, a Microsoft architect, devoted a famous online lecture for the programmers of his company, to stress the importance of an approach that is as simple as possible in developing IT solutions. The lecture was entitled "The

---

[35] In biology, phylogenetics is the field that refers to the inference on the evolutionary history within groups of organisms.

lost art of simplicity"[36] and it was constituted by a series of quotations and some nice cartoons. Among the many quotes reported by Josh Holmes, I report here two of them that I deem particularly significant in the present context. The first one is attributed to Albert Einstein and is absolutely enlightening considering the world of Big Data: "Every intelligent madman can make things bigger, more complex and more violent. It takes a touch of genius - and a lot of courage - to move in the opposite direction". The second is due to Mark Twain and, since I read it, it always comes to my mind every time I write an email: "I apologize for the length of this letter, but I didn't have time to make it shorter".

This last sentence would be the perfect expression that I could use to apologize to the reader for failing to further simplify a topic as complex as that of Big Data, given that the purpose explicitly stated from the beginning in the Introduction of this book was exactly this one. It would therefore be an ideal way to conclude this essay and take my leave of the reader if it was not because we need to treat, albeit briefly, another very important aspect of Big Data: the one concerning the ethical and privacy issues raised by the collection and use of an incredibly large number of data and individual confidential information.

This will be the purpose of the third and final chapter.

[36] Josh Holmes (2009) "The lost art of simplicity" https://www.slideshare.net/joshholmes/the-lost-art-of-simplicity.

Chapter 4
Big Data: The *Sixth Power*

The separation of powers is the fundamental principle of the rule of law theorized by the French judge Charles-Louis de Secondat, Baron de La Brède et de Montesquieu known by everybody simply as Montesquieu.

In his essay, written in 1748 "The spirit of the laws",[1] Montesquieu clearly explains that the principle of the separation of powers consists in the identification of three distinct and independent functions within the State and in the attribution to them of three distinct mutually balancing powers: The *legislative power*, which deals with creating laws, the *executive power*, which deals with applying them, and the *judiciary power,* which takes care of enforcing them. According to this principle, the three powers must be absolutely independent in order to guarantee democracy. As Montesquieu states:

> Everything would be lost if the same man, or the same body of elders, or nobles, or people, exercised these three powers: that of making laws, of making public decisions, and of judging crimes or private disputes.[2]

Following the early lesson of Montesquieu, in the common language, two more powers were added to this traditional subdivision in order to emphasize the importance of further elements which emerged as important over the years in democratic life. As reported by Thomas Carlyle, the term *Fourth Power* was first coined by the deputy Edmund Burke in a session of the House of Commons of the English Parliament in 1787 to refer to the press as an instrument of democratic life. In his book *On Heroes and Hero Worship* (Carlyle, 1840), Carlyle reports: "Burke said there were three Estates in Parliament; but, in the Reporters' Gallery yonder, there sat a Fourth Estate more important far than they all".

The term *Fifth Power*, on the other hand, has a more recent history and it derives from the name of an underground newspaper of the American counterculture of the

[1] Montesquieu (1748).

[2] Montesquieu (1748).

G. Arbia, *Statistics, New Empiricism and Society in the Era of Big Data*,
SpringerBriefs in Statistics,
https://doi.org/10.1007/978-3-030-73030-7_4

1960s, which was published in Detroit in 1965. The latter refers to the means of radio and television communication and, by subsequent extension to the Internet, as powerful instruments of control and possible manipulation of public opinion. In fact, the media guarantee the publicity of political life and, by doing so, they exercise an important function of balancing powers in the democratic system, if they remain clearly separated from the other three powers of the State. We have had in the past, and we still have, examples of the dangers which we may incur when press, radio and television are not sufficiently independent of the political power, thus undermining the democratic principles enshrined in Montesquieu and favouring totalitarian regimes.[3]

Those who can access Big Data and are able to implement the statistical methods necessary to extract useful information from them possess nowadays an important *Sixth Power*.[4] Like the other five powers mentioned before, also the *Sixth Power* has an important function of democratic rebalancing between the different forces by providing essential data to make decisions. However, this is true only if it obeys the Montesquieu's principle of a total independence, failing which it runs the risk of weakening (if not total annihilation) of democracy. However, the Sixth Power is more hidden than the others, even more subtle and pervasive than the great persuasive power of the press, television and the Internet. In fact, we can imagine (at least theoretically) to be able to escape the persuasive power of press and television, but we can never think of escaping the power of Big Data because of the infinite channels through which we unknowingly feed its expansion in every moment of our day. Indeed, we have already remarked many times in this essay how, in everyday life, we continuously transmit (consciously or not) messages about our preferences, our habits, our convictions and our choices. We do this explicitly when we use a social network, but unknowingly just by visiting a website, reading news, making a purchase and in many other situations that we incur routinely during the day.

Let us imagine that there exists someone who is able to collect all these data into a single scheme, merging different sources and thus putting together, using sophisticated statistical methods, the many pieces of the enormous puzzle constituted by our activities, attitudes and preferences. If we multiply this information for more than 7 billion Internet users in the world,[5] we derive the possibility of building a very accurate profile of all individuals and the power to interfere with their choices by sending messages targeted to their preferences using the infinite communication channels offered by the Internet. This is what already happens, at least partially, in the field of digital advertising where our preferences, coming from the univocal IP address[6] of our computer, are recorded and processed with statistical procedures

---

[3]Given its importance, the independence of the press and radio and television is a fundamental element in the calculation, for example, of the *Index of democracy* calculated by the Economist, which classifies all countries in different categories of democracies. https://www.eiu.com/topic/democracy-index.

[4]This locution was used for the first time by Bauman and Lyon (2014) albeit with a slightly different meaning from the one we use here.

[5]According to an estimate of Internet World Stats, the world Internet users in March 2020 were 7,796,949,710. See https://www.internetworldstats.com/stats.htm.

[6]See footnote 1 in Chapter 2.

and used to provide us with suggestions to direct our purchase choices. This is the method used, for example, by the service offered by Amazon to suggest books that we may like based on our previous choices or by social networks, such as Facebook or LinkedIn, when they suggest people that we can know, or from the tourist services provided by Booking, Expedia, Tripadvisor or Trivago, to suggest trips that we might like based on those made in the past or even just the websites we visited previously. Nothing strange: these services are indeed very useful to guide our choices. However, if such a methodology for the use of Big Data is extended to all our actions and if it is used to influence the democratic life of a country without maintaining the necessary third-party status compared to the other powers of the State, it could constitute a very serious danger for democracy.

Some recent examples have clearly shown the risks we can incur if the principle of independence is violated. In this sense, it is paradigmatic the case of Cambridge Analytica which the European data protection supervisor Emanuele Menietti defined as "the possible scandal of the century".

Let me summarize briefly the main facts of this famous controversy and its relevance for our study.

In the spring of 2018, the *Guardian* and the *New York Times* published a series of articles which claimed that the US online marketing company Cambridge Analytica used a huge amount of data taken from Facebook to condition the 2016 US presidential campaign and the referendum on the exit of the United Kingdom from the European Union (Brexit). Cambridge Analytica was a company specialized in gathering Big Data related to their users, monitoring their social network activities in order to identify, through appropriate statistical procedures, what is called the *psychometric profile* of each user. In addition to this information, Cambridge Analytica acquired many other data, collected by other companies, related to the traces that people left on the Internet which, combined with the previous ones, allowed them to develop a «behavioural microtargeting system» which, in turn, could be translated into highly personalized advertisements based on the tastes and emotions of each individual. The creator of these algorithms was Michael Kosinski, now professor of Organizational Behaviour at Stanford University. Speaking about his behavioural microtargeting system, Kosinski argued that by analysing only 10 of an individual's *likes*, it is possible to accurately predict the personality of a subject better than her colleagues. 70 likes would be enough to know more about people personality than her friends, 150 are enough to know them better than her family and 300 likes are enough to overcome her partner's knowledge. Kosinski also argues, provocatively, that with higher amounts of information it would be possible to know more about the personality of the subject than the subject knows himself/herself! According to the *Guardian* and to the *New York Times*, in the past, the information gathered by Cambridge Analytica would have been analysed jointly with those from an application called "*thisisyourdigitallife*" with which about 270,000 people connected via Facebook in order to receive their psychological profile obtaining in exchange some personal information together with the sharing (prior an explicit consent) of information related to their friends. This practice was soon blocked by Facebook because it was considered too invasive of people's privacy, but in that moment, the application had already collected data on Facebook users in a number which the New York Times and Guardian estimated to 50 million individuals. Sharing such data with Cambridge Analytica is in clear breach of Facebook's terms of use, which explicitly prohibit them from being shared with third-party companies.

Following the *Guardian* and *New York Times* investigation, in the spring of 2018, the special prosecutor investigating the alleged interference in the 2016 US presidential election requested Cambridge Analytica to provide documents on its own activities to dispel the suspicion that the company had used the data to make electoral propaganda favouring one of the candidates. In the same period, the *Guardian* also devoted a long investigation to the role of Cambridge Analytica in the Brexit referendum campaign, claiming that the company had collaborated in collecting data and information on users and then making propaganda in favour of the exit of the United Kingdom from the European Union.

In his defence, Cambridge Analytica stated that no rules had been broken in gathering and sharing data that were collected only with the consent of the users and having used them with methods that were already tested in some previous US presidential campaigns.

To clarify the main questions linked with the scandal, the United States Senate Judiciary Committee called a first hearing in April 2018, to investigate, in particular, Facebook's role in the breach and in the data privacy violation. During his testimony, the CEO of Facebook, the celebrated Mark Zuckerberg, admitted that the controls made by his platform were still insufficient and he publicly apologized for the lack of supervision in diffusing on the platform fake news and hate speeches and for not controlling foreign interference in the elections. He also admitted that in 2013 a data scientist of Cambridge Analytica, Aleksandr Kogan, created a personality quiz app (downloaded by about 300,000 users) that was able to retrieve Facebook information, even if it was only in 2015 that he was made aware of this. In any case, Zuckerberg agreed that it was inevitable to introduce new and more stringent rules for the use of data collected in the social networks.

The Congress also called a second hearing in May 2018, where the director of the Public Utility Research Center at the University of Florida, Dr. Mark Jamison, clarified that it was not unusual for presidential campaigns the use of Big Data to profile voters and that this happened already with the elections of Barack Obama and of George W. Bush campaign in the past. However, he also criticized Facebook for not being "clear and candid with its users" because the users were not aware of the extent that their data would be used.[7] In the same hearing, Christopher Wylie, a former Cambridge Analytica's director, said that he decided to denounce the breach because he wanted to "protect democratic institutions from rogue actors and hostile foreign interference, as well as ensure the safety of Americans online".[8] He also

---

[7]See "Mark Jamison's Written Testimony to the Senate". https://www.judiciary.senate.gov/imo/media/doc/05-16-18%20Jamison%20Testimony.pdf.

[8]See "United States Senate Committee on the Judiciary". https://www.judiciary.senate.gov/download/05-16-18-wylie-testimony. May 8, 2020.

explained how Cambridge Analytica used Facebook's data to categorize people and revealed the Russian contacts with Cambridge Analytica.

The impact of the scandal on the social network platform was very strong in a first moment, but also limited in time. In fact, in the first weeks after the Cambridge Analytica scandal, the number of likes and posts appearing on the site decreased by almost 20%, but the number of users kept on increasing in the same period. Facebook also suffered from an immediate financial loss with the stock falling down by 24% (about $134 billion) in few weeks after the scandal, but it took only a couple of months for the company to fully recover the losses. A number of campaigns were also initiated to boycott Facebook as a reaction to the data privacy breach, such as #DeleteFacebook and #OwnYourData, but Facebook did not notice "a meaningful number of people act" abandoning the social network.[9]

The dispute had a long history. In July 2018, the UK's Information Commissioner's Office (ICO) announced the willingness to fine Facebook £500,000 over the data breach. A year after, in July 2019, the Federal Trade Commission approved a fine of $5 billion to finally settle the investigation into the data breach. Other countries have also acted against Facebook for improper data use. For instance, Italy imposed initially a fine of 52,000 Euros for the violations of the missing consent of use. Then, the fine was raised to 1 million Euros in June 2019 due to the dimension of the database involved, to the economic conditions of Facebook and of the number of Italian users.

After a period when the problem of privacy related to the use of social media received comparatively relative less attention at least by the vast public, it was due in the end of 2020 that the dangers connected to the uncontrolled Sixth Power came again to the fore in the public opinion. And it was again connected to a US presidential election.

In fact, in the days immediately before the inauguration of Joe Biden as the new president of the United States, a violent riot and attack against the Congress were carried out by a group of supporters of Donald Trump aiming at overturning his defeat in the 2020 presidential election. This event is reported as the «*storming of the United States Capitol*». Breaching police perimeters, rioters vandalized parts of the building occupying it for several hours. In the aftermath of these dramatic events, several social platforms (namely, Twitter, Facebook and Instagram) temporarily blocked Trump on the ground that he used the social networks to cast doubts on the integrity of the election, to urge his supporters to go to Washington and to protest. The managers of these social networks interpreted Trump's messages as incentive to violence and, on this basis, decided to ban him from the networks in order to prevent further abuses. However, this decision cast new doubts about the control exerted by the Sixth Power and about its potential interference with the other democratic powers violating Montesquieu principle of separation. Many commentators, both among those that supported Trump and those opposing him, questioned if the decision to

[9]Guynn, Jessica. "Delete Facebook? It's a lot more complicated than that". USA TODAY. April 13, 2020.

allow or prohibit publishing on a social network could be left to the subjectivity of few private individuals and saw this act as a violation of the right of expression.

The question concerns who is entitled to exert a control on what is made publicly available on the social networks. Even if we consider it legitimate to control the opinions expressed to the public, who must implement such a control? Should it be a private or a public enforcement?

It can also be remarked that, even if a social platform belongs to a private company, its action affects public ethics so that it is only a public power that could regulate who on social media can express themselves and in what terms.

The controversial events described here had the merit of bringing to the core of the debate new elements related to the use of Big Data, making the public aware of the risks associated with them and the need for more restrictive regulations on their unauthorized use due to the danger of interference with political power in violation of the principle of separation. The problem obviously concerns all online companies that offer their services and collect information about users. More precise regulations are certainly necessary and have been urged for some time by online privacy protection organizations. In recent years, the European Union has tightened up these rules, but given the global nature of the problem, it is only from coordinated action by all countries that the problem can be satisfactorily resolved so as to guarantee the ethical use of Big Data. The events of 2021 described here show that, when this book is printed, we are still far from finding a solution which is widely agreed.

Chapter 5
Conclusions: Towards a New Empiricism?

The last decades have seen a formidable explosion in data collection and in their diffusion and use in all sectors of human society. This phenomenon is mainly due to the increased ability to collect and store information automatically through very different sources such as sensors of various kinds, satellites, mobile phones, Internet, drones and many others.

In this essay, my aim was to describe, using an informal approach accessible to non-specialists in the field, the main characteristics and problems related to this phenomenon, called Big Data, and its possible consequences in everyday life.

As Michael Coren states in the quote that opens this essay, "in every century new technology has wiped out the old world with the vision of a new one". This happened for the steam engine, for electricity, for atomic energy and for microprocessors and is now happening for Big Data.

The collection and dissemination of enormous amounts of data certainly have the potential to revolutionize the way decisions are made in all fields, as companies, as public operators and as a single individual. Indeed, it is becoming increasingly clear that companies and public administrations will have to be able to base their decisions on empirical data in order to be competitive. In the same way, each of us, to make small or big decisions, will increasingly make use of the empirical data widely available from very different sources. In order to do this, both individuals and institutions will have to acquire new skills and competences so as to be able to draw from the vast mass of available data, information that is useful to build up knowledge. It is out of the question that, at the moment, we are still not ready to face such a revolution. As stated by the aforementioned Hal Varian "Data is widely available. It is the ability to extract wisdom from them that is scarce".

Moreover, when an epochal revolution explodes, it does not ask permission first, does not knock on the door and does not give advance notice so that everyone can prepare themselves properly. It just explodes. The introduction of the steam engine or electricity took years before its widespread distribution. Similarly, the first microprocessor was produced in the 1960s, but it took decades before its use

G. Arbia, *Statistics, New Empiricism and Society in the Era of Big Data*,
SpringerBriefs in Statistics,
https://doi.org/10.1007/978-3-030-73030-7_5

benefited a vast audience. As with any other revolution, those who adapt quicker to the Big Data revolution will be able to benefit the most.

To do this, each of us, in public and in private sectors, will have to progressively acquire a more empirical-inductive mentality in sharp contrast to the logical-deductive approach which is still prevalent nowadays in many fields. Each of us will have to analyse the choices, the decisions and the various alternatives that gradually appear, with an attitude which should be similar to that of a scientist who analyses the results of an experiment.

From this viewpoint, we are at the dawn of a new social-empiricism, whose outlines and possible distortions are already visible.

It will involve a change of mentality that will probably take a few generations.

To moderate the euphoria towards the advancing new world, however, it has been observed that the accelerated (and often uncontrolled) development of Big Data influences the privacy of individuals and makes them vulnerable to unauthorized access to their personal and influential information. In the same way, as we have tried to illustrate in Chapter 4, individual freedom is threatened by the pervasiveness and persuasiveness of the power of Big Data through their improper use with the intent to direct our economic and political choices. In this sense, I hope to have been successful in illustrating the fact that the Big Data phenomenon does not only concern statistical, economic or information technologies, but it also has important ethical-legal implications.

Just as the attitude towards empirical data will progressively change in each of our decisions, at the same time it will be necessary to develop a critical approach towards them considering that they are not always a reliable and distortion-free source of information.

It seems legitimate to be able to affirm that in this emerges a new role of Statistics, a discipline so far conceived as the art of knowing reality and making decisions based on empirical-observational data. In fact, while in the past Statistics has essentially played the role of suggesting rigorous methods for data collection and analysis both in the descriptive and in the inferential-inductive area, in a society in which huge masses of data are produced every day from very different (and often uncontrolled) sources, Statistics will increasingly have to devote itself to developing techniques aimed at validating the data and corroborating and certifying the goodness of their sources.

In this, a fundamental role should be played by the national and supranational statistical institutes, which will have to reposition themselves in a global information market, proposing themselves less and less as exclusive data producers and increasingly as guarantors of the quality of data which are produced by alternative sources as well as of their reliability.

Only in this case, we will not run the risk of finding ourselves lost in the corridors of the immense Babel library that is taking shape around us, enticed by the reading of a myriad of deceptive and liar volumes and without having the slightest possibility of identifying which book we are really interested in.

Correction to: Statistics and Empirical Knowledge

Correction to:
Chapter 3 in: G. Arbia, *Statistics, New Empiricism and Society in the Era of Big Data*, SpringerBriefs in Statistics, https://doi.org/10.1007/978-3-030-73030-7_3

The original version of this chapter was inadvertently published with two incorrect Figures 3.1 and 3.2, which have now been updated. The book has been updated with the changes.

The updated version of this chapter can be found at
https://doi.org/10.1007/978-3-030-73030-7_3

G. Arbia, *Statistics, New Empiricism and Society in the Era of Big Data*,
SpringerBriefs in Statistics,
https://doi.org/10.1007/978-3-030-73030-7_6

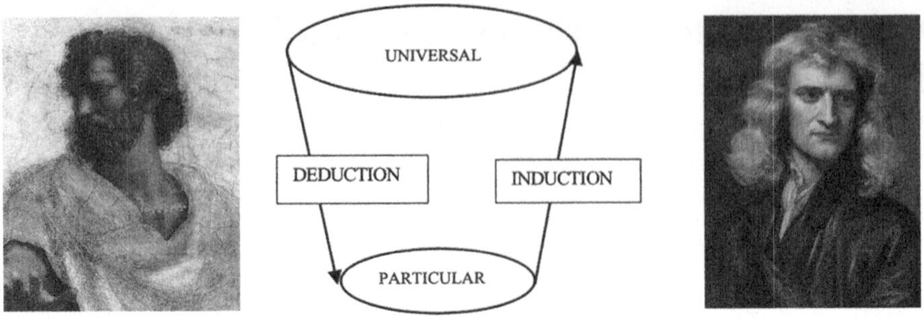

Fig. 3.1 Two basic procedures to reach knowledge: Deduction and Induction

Fig. 3.2 Deduction and induction in the statistical terminology

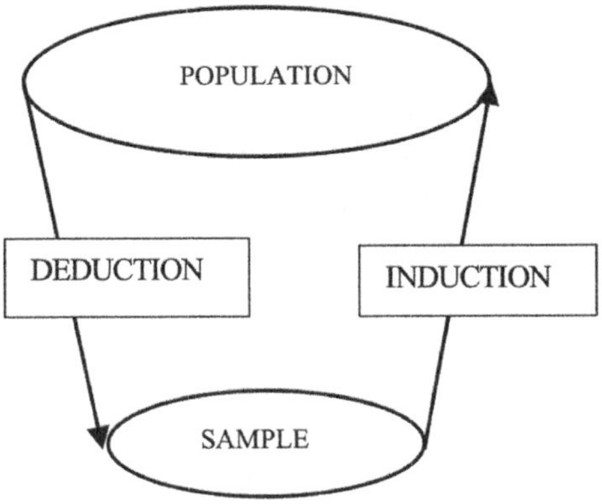

References

Arbia, G. (2018). *Statistica, nuovo empirismo e società nell'era dei Big Data.* Nuova Cultura.

Arbia, G., Genovese, G., Micale, F., Nardelli, V., & Solano-Hermosilla, G. (2020). *Post-sampling crowdsourced data to allow reliable statistical inference: The case of food price indices in Nigeria.* arXiv:2003.12542, 27 March 2020.

Aristotel. (1933). *Metaphysics,* (written IV century BC) (Hugh Tredennick, Trans., 2 vols. Loeb Classical Library 271, 287). Harvard University Press.

Aristotel (1989). *Topica* (written IV century BC) (E. S. Forster, Trans., Loeb Classical Library, Book I, 12, 105 a 11). Cambridge: Harvard University Press.

Azzalini, A. (1996). *Statistical Inference: Based on the likelihood*. Chapman & Hall.

Bachelard, G. (1934). *Le Nouvel Esprit Scientifique, English Edition, The new scientific spirit* (1985, A. Goldhammer, Trans.). Boston: Beacon Press.

Bassi, F., Arbia, G., & Falorsi P. D. (2020). *Observed and estimated prevalence of Covid-19 in Italy: How to estimate the total cases from medical swabs data, Science of The Total Environment*. ISSN 0048-9697. https://doi.org/10.1016/j.scitotenv.2020.142799.

Bauman, Z., & Lyon, D. (2013). *Liquid surveillance.* Cambridge: Polity Press.

Borges J. L. (1999). *Collected fictions* (Andrew Hurley, Trans.). Penguin Classics Deluxe Edition.

Carlyle, T. (1840). Lecture V: The hero as man of letters. Johnson, Rousseau, Burns. In *On heroes, hero-worship, & the heroic in history. Six lectures. Reported with emendations and additions* (Dent, 1908 ed.). London: James Fraser.

Davenport, T. H. (2014). *Big data at work.* Harvard Business Review Press.

Finger, L. (2014). *Recommendation engines: The reason why we love big data.* MIT Management Executive Education.

Fisher, R. A. (1922). On the mathematical foundations of theoretical statistics. *Philosophical Transactions of the Royal Society a, 222,* 309–368.

Fisher, R. A. (1925). *Statistical methods for research workers.* Edinburgh: Oliver and Boyd.

Fisher, R. A. (1935). *The design of experiments.* New York: Hafner Publishing Company.

Franks, B. (2012). *Taming the big data tidal wave.* Hoboken, NJ: Wiley.

Kant, I. (1787). *Critique of pure reason* (2nd ed., Norman Kemp Smith Editor, published in 2007). Palgrave Macmillan.

Kuhn, Thomas S. (1962). The structure of scientific revolutions (3rd ed., 1996). Chicago, IL: University of Chicago Press.

Kundera, M. (2004). *The unbearable lightness of being* (Michael Henri Heim, Trans.). Harper-Collins.

G. Arbia, *Statistics, New Empiricism and Society in the Era of Big Data,*
SpringerBriefs in Statistics,
https://doi.org/10.1007/978-3-030-73030-7

Laney, D. (2001).*3-D data management: Controlling data volume, velocity and variety*, in Application Delivery Strategies by META Group Inc., p. 949. https://blogs.gartner.com/doug-laney/files/2012/01/ad949-3D-Data-Management-Controlling-Data-Volume-Velocity-and-Variety.pdf .
Li, R., Pei, S., Chen, B., Song, Y., Zhang, T., Yang, W., & Sharmna, J. (2020). Substantial undocumented infection facilitates the rapid dissemination of novel coronavirus (SARS-CoV-2). *Science, 368, 6490,* 489–493.
McAfee, A., & Brynjolfsson, E. (2012). *Big data: The management revolution*. Harvard Business review.
McKinsey. (2011, June). *Big data: The next frontier for innovation, competition and productivity.* The McKinsey Global Institute.
McKinsey. (2013, March). *McKinsey quarterly Big Data, what is your plan?* By Stefan Biesdorf, David Court, and Paul Willmott.
Montesquieu, C. L. de (1748). *De l'esprit des loix* (English translation). In Anne M. Cohler, Basia Carolyn Miller, & Harold Samuel Stone (Eds.), *The Spirit of the Laws (2002)*. Cambridge Texts in the History of Political Thought. Cambridge: Cambridge University Press.
Moore, G. (2014). *The business book* (p. 316). Dorling Kindersley Ltd.
Newton, I. (1687). *Philosophiae naturalis principia mathematica*, Londini, iussu Societatis Regiae ac typis Josephi Streater, *Isaac Newton's Natural Philosophy* (English translation edited by Jed Z. Buchwald & I. Bernard Cohen). MIT Press.
Parker, C. B. (2015). Michal Kosinski: Computers are better judges of your personality than friends. *Operations, information & technology*. Stanford Graduate School of Business.
Petty, W. (1767). *Political arithmetic posthum* (written approx. 1676).
Poincarè J.-H. (1908). *Science and method* (Francis Maitland, Trans., Published in 2010). Cosimo classics.
Popper, K. R. (1959). *Logik der Forschung. Zur Erkenntnistheorie der modernen Naturwissenschaft* (English translation *The logic of scientific discovery* [2005]). Routledge Classics.
Schmidt, E. (2010). *Techonomy conference*. CA: Lake Tahoe.
Schoenberger, V. M., & Kenneth Cukier, K. (2013). *A revolution that will transform how we live, work and think: Big data*. London: John Murray.
Shannon, C. E. (1948). A mathematical theory of communication. *Bell System Technical Journal, 27,* 379–423 (july), 623–656 (october).
Shelling T. C. (1978). *Micromotives and Macrobehavior*. Norton.
Smith, V. L. (2008). Experimental methods in economics. In *The new Palgrave dictionary of economics* (2nd ed.).
Thiele, T. N. (1889). *Forelæsninger over almindelig Iagttagelseslære: Sandsynlighedsregning og mindste Kvadraters Methode*. Reitzel, København. The general theory of observations: Calculus of probability and the method of least squares translated by S. L. Lauritzen. Oxford University Press.
Van Fraassen, B. C. (1980). *The scientific image*. Oxford University Press.
Voltaire, J. J. (1734). *Lettres philosophiques*, n. 15. Philosophical Letters (Prudence Steiner, Trans., 2007). Hackett Publishing Company.
Volterra, V. (1920). *Sui tentativi di applicazione della matematica alle scienze biologiche e sociali.* Discorso inaugurale" in Annuari della Regia Università di Roma, 1901, 3–28. Some essays are translated in English and reported in Guerraggio, Angelo and Paoloni (2013) Vito Volterra.

The manufacturer's authorised representative in the EU is Springer Nature Customer Service Centre GmbH, Europaplatz 3, 69115 Heidelberg, Germany. If you have any concerns regarding our products, please contact ProductSafety@springernature.com

Printed and bound by CPI Group (UK) Ltd, Croydon, CR0 4YY
30/09/2026
02225708-0004